Basic Organic Chemistry for the Life Sciences

Hrvoj Vančik

Basic Organic Chemistry for the Life Sciences

 Springer

Hrvoj Vančik
Department of Chemistry
University of Zagreb
Zagreb, Croatia

ISBN 978-3-319-36156-7   ISBN 978-3-319-07605-8 (eBook)
DOI 10.1007/978-3-319-07605-8
Springer Cham Heidelberg New York Dordrecht London

© Springer International Publishing Switzerland 2014
Softcover reprint of the hardcover 1st edition 2014
This work is subject to copyright. All rights are reserved by the Publisher, whether the whole or part of
the material is concerned, specifically the rights of translation, reprinting, reuse of illustrations, recitation,
broadcasting, reproduction on microfilms or in any other physical way, and transmission or information
storage and retrieval, electronic adaptation, computer software, or by similar or dissimilar methodology
now known or hereafter developed. Exempted from this legal reservation are brief excerpts in connection
with reviews or scholarly analysis or material supplied specifically for the purpose of being entered and
executed on a computer system, for exclusive use by the purchaser of the work. Duplication of this
publication or parts thereof is permitted only under the provisions of the Copyright Law of the Publisher's
location, in its current version, and permission for use must always be obtained from Springer.
Permissions for use may be obtained through RightsLink at the Copyright Clearance Center. Violations
are liable to prosecution under the respective Copyright Law.
The use of general descriptive names, registered names, trademarks, service marks, etc. in this publication
does not imply, even in the absence of a specific statement, that such names are exempt from the relevant
protective laws and regulations and therefore free for general use.
While the advice and information in this book are believed to be true and accurate at the date of
publication, neither the authors nor the editors nor the publisher can accept any legal responsibility for
any errors or omissions that may be made. The publisher makes no warranty, express or implied, with
respect to the material contained herein.

Printed on acid-free paper

Springer is part of Springer Science+Business Media (www.springer.com)

This work is the result of my collaboration with plenty of my colleagues and students during more than twenty years of my lecturing of organic chemistry. The manuscript in this form would not be possible without the insightful comments of reviewers Mladen Mintas, Miroslav Bajić, Srđanka Tomić-Pisarović (University of Zagreb, Croatia) and Igor Novak (Charles Sturt University, Sydney, Australia), to whom I owe a debt of gratitude.

Preface

This textbook appears as a result of the experience in more than 20 years of lecturing organic chemistry to students of biology, molecular biology, and ecology within the Faculty of Science and Mathematics of the University of Zagreb. Since the great books of organic chemistry for chemists appear to be too advanced for students whose study is only partially related to chemistry, I have decided to prepare the text that is more oriented to the essence of organic chemistry.

Open problems in writing the basic organic chemistry textbook include the selection of concepts for the representation of the material, but also the level of the explanation of the complex phenomena such as reaction mechanisms or the electron structure. Here I propose the compromises. First compromise is related to the mode of the systematization of the contents, which can traditionally be based either on the classes of compounds, or on the classes of reactions. Here, the main chapter titles contain the reaction types, but the subtitles involve the compound classes. The electronic effects as well as the nature of the chemical bond is described by using the quasi-classical approach starting with the wave nature of the electron, and building the molecular orbitals from the linear combination of the atomic orbitals on the principle of the qualitative MO model. Hybridization is avoided because all the phenomena on this level can be simply explained by non-hybridized molecular orbitals.

The text is divided in two parts. First chapters deal with fundamental aspects of the structural theory, reaction dynamics of organic reactions, electronic structure, and some basic spectroscopy. Last, the largest chapter represents the introduction to the organic chemistry of natural products. Comparison of the reactions in the laboratory with the analogous molecular transformations in living cells will help the students to understand the basic principles of biochemistry. The most interesting property of organic chemical systems, the formation of the high diversity of structures, is pointed out almost in all chapters. This approach is designed to help the students to provide deeper insight into the phenomena of the chemical evolution as a base for the biological evolution.

I intend this book for students of biology, molecular biology, ecology, medicine, agriculture, forestry, and other professions where the knowledge of organic chemistry plays the important role. I also hope that the work could also be of interest to non-professionals, as well as to the high school teachers.

Zagreb, Croatia   Hrvoj Vančik
2014

Contents

Introduction

The historic moment when organic chemistry appeared as part of chemistry, where organic chemistry deals with compounds originating from living organisms, is difficult to establish. More than 200 years ago, in 1784, T. Bergman used the term *organic chemistry* for the first time. Perhaps, two historic events in the development of chemistry could be regarded as crucial for the development of this branch of science.

In the first place one should mention **Jön Jacob Berzelius**, who, at the beginning of the nineteenth century, developed the method for the systematic elemental analysis of organic substances. Berzelius observed that all organic substances produce carbon dioxide and water upon combustion. By accurately measuring the masses of these products he calculated the percentages of carbon, hydrogen and oxygen in organic compounds. The most important conclusion was that all organic compounds consist of carbon and hydrogen. Accordingly, organic chemistry could also be called the chemistry of carbon compounds.

Secondly, perhaps the most important event in the development of organic chemistry has been the discovery that an organic compound could be prepared from inorganic starting material. In 1828, **Friedrich Wöhler** successfully prepared urea, the organic compound, by heating the inorganic salt ammonium cyanate. Before this discovery, organic substances were thought to be exclusively derived from living organisms.

One of the fundamental and general questions about the nature of organic compounds is the special nature of carbon as the basic element from which all the organic substances and all the known substances of Life are built. Could silicon, for instance, the element which is in the same group of the periodic table as carbon, be the basic element for the development of some alternative life?

To answer this question it is necessary to look at the special properties of the element carbon, the properties which are responsible for the emergence and evolution of complex organic molecules. The most important prebiotic condition for the beginning of biological evolution is the appearance of a complex mixture of molecules with a high diversity of structures. **Stuart Kauffman** calculated that such critical diversity should comprise at least 200,000 molecules with different structures. Basic properties of carbon are such that compounds of this element can form an enormous number of structures. More than ten million organic compounds are known at present.

In the second half of the nineteenth century, **August Kekulé, Archibald Couper** and independently **Aleksandr Butlerov** discovered the most important property of carbon: the mutual binding of carbon atoms into chains, branched chains and ring structures. This discovery has been extended with the knowledge that carbon atoms can be connected by stable single, double and triple bonds. For comparison, silicon atoms can also be connected by single and double bonds, but in contrast to carbon-carbon bonds, silicon-silicon bonds are weaker, unstable and sensitive to light. Hence, silaorganic compounds would not be able to survive under the natural conditions which were present on Earth at the time when biological evolution started. However, silicon atoms do form strong bonds with oxygen atoms and form a high diversity of structures in which silicon-oxygen-silicon structural motifs are present. Such structures are characteristics of the terrestrial mineral world.

Later in this book, we will introduce a series of other special properties which make carbon a unique biogenic element.

Structures of organic molecules can be drawn from the structures in which carbon appears as an element i.e. from the structures of its allotropic modifications. Until the last quarter of the twentieth century only two allotropic modifications of carbon were known, **graphite** and **diamond**. From the viewpoint of structural organic chemistry, structures of graphite and diamond represent basic structural patterns by which carbon atoms can be interconnected.

Graphite   Diamond

In graphite, every carbon is surrounded by three neighboring carbon atoms in such a way that all four atoms lie in the same plane. In contrast, the carbon atoms in diamond are arranged in the three-dimensional array where every atom is surrounded by four neighbors, which are configured in tetrahedral geometry. These two motifs, tetrahedral and planar trigonal, respectively, represent basic structural patterns of organic molecules. Linear binding of atoms is also possible for organic molecules, but allotropic modification of carbon with such structure has not been observed yet.

Tetrahedral arrangement

Trigonal arrangement

In the intergalactic space there are stars which are in the last phase of their development and which produce a lot of elemental carbon by eruptions. **Harold Kroto** and his collaborators **Richard Smalley** and **Robert Curl** have investigated in detail the nature of such intergalactic carbon. The result of their research has been the discovery of a new allotropic modification of carbon in which atoms form structures resembling a ball. By measuring the relative molecular masses of such ball-molecules and simulating the interstellar conditions in the laboratory, Kroto, Smalley and Curl have found that these molecules mostly consist of 60 carbon atom clusters distributed as pentagonal and hexagonal structures. There are 12 pentagons surrounded by hexagons. Since the proposed structure resembles some works of art, especially the architecture constructed by the architect **Richard Buckminster Fuller**, this C_{60} molecule has been named **fullerene**. In subsequent research a series of similar ball-like structures was discovered, some of which also have tubular structures of carbon atoms. These molecules have dimensions on the nanometer scale and have intriguing properties which are interesting for use in the sophisticated technology of novel materials and electronics. The discovery of fullerenes represents the beginning of the new era of **nanotechnology**.

FULLERENE

Chapter 1
Alkanes, Composition, Constitution and Configuration

In principle, the organic molecules may be considered as consisting of a hydrocarbon skeleton to which functional groups are attached. While the hydrocarbon skeleton determines molecular shape and flexibility, the chemical reactivity depends mostly on the functional groups present. For a better understanding of the basic properties and structure of the molecular skeleton, let us start with hydrocarbons, which represent organic compounds without any functional groups.

As compounds which consist of carbon and hydrogen atoms only, the hydrocarbons can be divided in two main categories, **saturated**, which comprises alkanes and cycloalkanes, and unsaturated, which comprises of alkenes, alkynes, and aromatics. The term "saturated" indicates the impossibility of adding more hydrogens to the molecule.

In the following scheme, ethene as an unsaturated compound can be transformed to the saturated ethane compound via the process of binding a hydrogen molecule. Since all valencies of the carbon atoms in ethane are occupied there is no place to add any more hydrogen atoms.

Ethene   Ethane

Nature abounds in hydrocarbons especially in crude oil and natural gas. The mixture of hydrocarbons present in oil can be separated into groups of compounds with different boiling points through the industrial process of fractional distillation. By analysis of compounds from different fractions it is possible to elucidate their elemental **composition** from which in turn, their chemical formulas can be calculated, for instance C_2H_6 for ethane. The determination of composition is based on the property of hydrocarbons to combust into water and carbon dioxide. At the beginning of the

H. Vančik, *Basic Organic Chemistry for the Life Sciences*,
DOI 10.1007/978-3-319-07605-8_1, © Springer International Publishing Switzerland 2014

nineteenth century, **Jön Jacob Berzelius** calculated formulas for a series of organic compounds from the measured masses of water and carbon dioxide.

Although knowledge of the composition is very important for the classification of organic compounds, for the description of organic molecules in more detail it was nonetheless insufficient. Difficulties arose because many different organic compounds have the same composition. Atomic theory that could help resolve these contradictions was still in the early stages of development at the time. **John Dalton** had published his discovery of atoms only in 1808 in his book *The New System of Chemical Philosophy*.

An additional layer of controversy appeared during this time. It was believed that inorganic and organic compounds have different natural origins. Hence it was thought impossible that organic compounds could be prepared from any inorganic source. They, so it was thought, could be obtained exclusively from living organisms. This *vitalistic* point of view was disproved by a student of Berzelius, **Friedrich Wöhler**, who succeeded in preparing an organic compound from an inorganic precursor. Wöhler's organic compound urea, was obtained by heating the inorganic salt ammonium cyanate:

$$\overset{+}{N}H_4\overset{-}{O}CN \xrightarrow{\Delta} H_2NCONH_2$$

**Ammonium
cyanate**   **Urea**

Composition CH$_4$N$_2$O

(symbol Δ is for heating)

Wöhler's experiment is important not only because of its finding that there is only one chemistry, independent of the origin of the substances, inorganic or biological, but because it demonstrated that two very different substances can have the same composition, in this case CH$_4$N$_2$O. The idea of **structure** as a higher level of organizing principle of matter has emerged. **Charles-Frédéric Gerhardt**, **August von Hofmann** and **Alexander Williamson** have around 1850 developed this idea into the concept of **constitution**, which describes the way in which atoms are interconnected in the molecule. In Wöhler's experiment different *constitutions* of ammonium cyanate and urea can be described by different *structural formulas*:

**Ammonium
cyanate**   **Urea**

Compounds such as urea and ammonium cyanate, which have the same composition but different constitution are called **isomers**. After this first example it has been found that there is an enormous number of isomers of organic compounds, especially hydrocarbons, in nature. Before clarifying these concepts in more detail let us describe the composition, constitution and names of some of the most important saturated hydrocarbons, **alkanes**. As can be seen from the following Table, all the given chemical formulas can be reduced to the general formula C_nH_{2n+2}, where n is an integer. Such alkanes are called n-alkanes and belong to a **homologous series** of compounds.

Different isomers are possible if the alkane molecule contains more than three carbon atoms. In the case of butane, which has the composition defined by the formula C_4H_{10}, the carbon atoms can be interconnected in two different ways forming two constitutional isomers. While the compound with the linear carbon chain is usually called n-**butane** its branched isomer is called *iso*-**butane**.

Structures shown in the following scheme are two different isomers which really represent two different compounds. To convert one isomer into another it is necessary to break and reform chemical bonds by a chemical reaction. The isomers are represented in two ways, by connectivity structural formula in which all the interatomic bonds and atomic symbols are shown and also by the condensed formula where the chains between the CH_3 and CH_2 groups are drawn. Such more practical formulas are mostly used in organic chemistry. The particular groups in the chain have special names such as **methylene group** for CH_2 and **methyl group** for CH_3.

Let us consider the number of isomers of alkanes for different numbers of carbon atoms. For instance, pentane can form three isomers and hexane has five isomers:

$$CH_3CH_2CH_2CH_2CH_3 \qquad CH_3\overset{\overset{\displaystyle CH_3}{|}}{C}HCH_2CH_3 \qquad H_3C\!-\!\overset{\overset{\displaystyle CH_3}{|}}{\underset{\underset{\displaystyle CH_3}{|}}{C}}\!-\!CH_3$$

ISOMERS of PENTANE

$$CH_3CH_2CH_2CH_2CH_2CH_3 \qquad CH_3CH_2\overset{\overset{\displaystyle CH_3}{|}}{\underset{\underset{\displaystyle CH_3}{|}}{C}}HCH_2CH_3 \qquad CH_3\overset{\overset{\displaystyle CH_3}{|}}{C}HCH_2CH_2CH_3$$

ISOMERS of HEXANE

In following table, the number of isomers is given for the alkanes with up to 10 C-atoms. The table shows that the number of isomers increases exponentially with the number of carbon atoms in the alkane molecule. The saturated hydrocarbon with 20 carbon atoms with the brutto formula $C_{20}H_{42}$ can have 366,319 isomers!

Number of C-atoms in alkanes	Number of possible constitutional isomers
1	1
2	1
3	1
4	2
5	3
6	5
7	9
8	18
9	35
10	75

Such a large number of isomers of simple alkanes explains why carbon as an element is unique and why it serves as a basis for the vast diversity of structures necessary for Life. This propensity for generating great diversity of organic molecules starting from simple structures will be exemplified further in later parts of this book.

Regarding isomers, it must be noted that depending on the way the atoms are interconnected, C-atoms can bind to each other in several different ways. Some carbons are bound to only one neighboring C-atom, some to two, three or four. Based on this criterion the carbon atoms are named **primary, secondary, tertiary** and **quaternary**, respectively:

Secondary C-atom, Primary C-atom, Tertiary C-atom, Quaternary C-atom

$$CH_3CH_2CH_2CH_2CH_2CH_3 \qquad CH_3CH_2CHCH_2CH_3 \qquad H_3C-\underset{\underset{CH_3}{|}}{\overset{\overset{CH_3}{|}}{C}}-CH_2CH_3$$

1.1 On the Nomenclature of Organic Compounds

When discussing chemistry as a discipline we must be aware of three different categories which are present in the methodology of chemical science and practice. Chemistry can be recognized by its **content**, **writing** and **language**. The most important is the content; these are the substances with which we have immediate experience either in laboratory or in everyday life. Chemical writing and language are human inventions; they are a kind of models which serve for more or less unambiguous communication between chemists about substances, chemical concepts and theories. While chemical writing comprises formulas which we have already mentioned, by chemical language we describe constitutions and configurations of molecules.

The chemical language is designed to be sufficiently precise so that from the name of a compound only one structural formula can be deduced. The names of compounds are based on the linguistic rules called **nomenclature**. Today, the chemical nomenclature is universal, standardized and governed by international conventions promulgated by the International Union for Pure and Applied Chemistry (IUPAC). According to IUPAC convention the name of a compound derives from the root of the word to which the prefixes and suffixes can be added, depending on the class and structure of the molecule. The root of the name is based on the number of C-atoms in the longest carbon chain and is derived from the names of simple hydrocarbons.

The suffix labels the functional group whose presence places the molecule into the appropriate class of chemical compounds. In this scheme the saturated hydrocarbons, the **alkanes** have the suffix **-ane**. For naming isomers, the system is more complicated and includes additional rules. Since the molecules of isomers are branched, the root name must correspond to the longest chain. The sidechains are treated as additional groups called **substituents**. In the final name of the structure, the substituents are introduced as prefixes to the root. The names of substituents are formed following the same rules as in the case of simple alkanes, i.e. the number of C-atoms followed by the suffix **-yl**.

Number of C-atoms	Formula	Root	Suffix	Name of compound	Substituent	Suffix	Name
1	CH_4	met-	-ane	methane	$-CH_3$	-yl	methyl
2	C_2H_6	et-	-ane	ethane	$-C_2H_5$	-yl	ethyl
3	C_3H_8	prop-	-ane	propane	$-C_3H_7$	-yl	propyl
4	C_4H_{10}	but-	-ane	butane	$-C_4H_9$	-yl	butyl
5	C_5H_{12}	pent-	-ane	pentane	$-C_5H_{11}$	-yl	pentyl

(continued)

(continued)

Number of C-atoms	Formula	Root	Suffix	Name of compound	Substituent	Suffix	Name
6	C_6H_{14}	hex-	-ane	hexane	$-C_6H_{13}$	-yl	hexyl
7	C_7H_{16}	hept-	-ane	heptane	$-C_7H_{15}$	-yl	heptyl
8	C_8H_{18}	oct-	-ane	octane	$-C_8H_{17}$	-yl	octyl
9	C_9H_{20}	non-	-ane	nonane	$-C_9H_{19}$	-yl	nonyl
10	$C_{10}H_{22}$	dec-	-ane	decane	$-C_{10}H_{21}$	-yl	decyl

The position of the substituent on the longest chain is labeled by a number. Number 1 must be assigned to the terminal carbon of the chain so that all other substituents can be labelled by the smallest possible numbers. If on the same basic chain two or more identical substituents are attached, the suffix is expanded by adding labels di-, tri-, etc. For instance, in the name 2,2-dimethylpropane (shown in the schemes below), the numbering 2,2- means that two methyl groups are bound to carbon 2.

$$CH_3$$
$$|$$
$$CH_3CHCH_2CH_3$$
$$1 \quad 2 \quad 3 \quad 4$$

2-Methylbutane

$$CH_3$$
$$|$$
$$CH_3CHCH_2CH_3$$
$$4 \quad 3 \quad 2 \quad 1$$

3-Methylbutane

$$CH_3$$
$$|$$
$$CH_3CHCH_2CH_2CH_3$$
$$1 \quad 2 \quad 3 \quad 4 \quad 5$$

2-Methylpentane

correct!

$$CH_3$$
$$|$$
$$CH_3CHCH_2CH_2CH_3$$
$$5 \quad 4 \quad 3 \quad 2 \quad 1$$

4-Methylpentane

incorrect!

$$CH_3$$
$$|$$
$$H_3C-C-CH_3$$
$$1 \quad |2 \quad 3$$
$$CH_3$$

2,2-Dimethylpropane

$$5 \quad 4 \quad 3 \quad 2 \quad 1$$
$$CH_3CH_2CHCH_2CH_3$$
$$|$$
$$CH_3$$

3-Methylpentane

$$H_3C \quad \quad 1 \quad CH_3$$
$$\diagdown \quad 2 \quad \diagup$$
$$CH-CH$$
$$H_3C \quad 3 \quad \diagdown CH_3$$
$$4$$

2,3-Dimethylbutane

$$CH_3$$
$$|$$
$$H_3C-C-CH_2CH_3$$
$$1 \quad |2 \quad 3 \quad 4$$
$$CH_3$$

2,2-Dimethylbutane

1.2 Configurations and Shapes of Molecules

The basic idea that molecules are real particles which have particular shape originated from three chemists, one of them was organic, the other inorganic and the third one was a physical chemist. By studying the symmetry of crystals of the organic salt ammonium-sodium tartarate, which had been isolated from the reaction mixture in alcoholic fermentation, **Charles LeBel** who worked with **Louis Pasteur** proposed in 1874 that the atoms bound to the central carbon atom in substituted alkanes are distributed in space so as to form a tetrahedron. Such tetrahedral spatial configuration resembles the distribution of C-atoms in diamond. Details of this discovery by LeBel will be discussed later in this book. The same idea about the tetrahedral structure of the alkane-like molecules has been independently proposed by physical chemist **Jacobus van't Hoff**, who studied isomers of substituted alkanes. The concept of spatial structure of inorganic compounds in which the atoms surrounding the central metal atom form an octahedron, was proposed by the inorganic chemist **Alfred Werner**.

The methane molecule CH_4, has the shape of a tetrahedron in which the carbon atom is at the center. Spatial three dimensional distribution of atoms in a molecule is called **configuration** and we can say that the methane molecule has a tetrahedral configuration. For pictoral representation of such spatial distribution there is a convention such that chemical bonds which lie in the plane of the drawing are labeled with a full line, the bonds located above the plane of the drawing by a wedge (bold elongated triangle) and the bonds below the plane with a dashed line (dashed elongated triangle). The angle between two C-H bonds $109°28'$, is known as the tetrahedral angle.

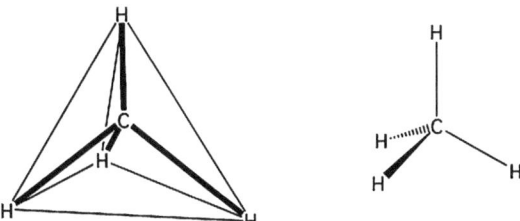

The fact that the molecule has such a distinct geometric form can be explained by the branch of physics known as **quantum mechanics**. In other words, the tetrahedral configuration of C-H bonds is the consequence of the repulsion of electron pairs which tend to be as far apart as possible from each other. This method of prediction of the molecular shape by considering the optimal distribution of bonds (bonding electron pairs) in which the electron repulsion is minimal, is called **VSEPR** (**v**alence **s**hell **e**lectron **p**air **r**epulsion). Although this method is widely used, in practice we must point out that this procedure is a simplified approach that can afford only an approximate picture of the molecule.

Let us use the knowledge about the tetrahedral shape to build structures of other simple alkane molecules. We observe that the consequence of tetrahedral structure is the zig-zag form of the alkane chains. Chemical formulas in the following scheme are called wedge-dash formulas.

Bearing this shape in mind, it is possible to write the structural formula in an even simpler form. The figure below shows the structures of alkanes, their simplified structural formulas as well as their condensed structural formulas.

$CH_3CH_2CH_3$ $CH_3CH_2CH_2CH_3$ $CH_3CH_2CH_2CH_2CH_3$

Although simplified, this geometry model explains molecular shapes satisfactorily enough to afford an approximate picture of molecules. More detailed insights into the molecular shapes is possible by using special microscopy techniques called **scanning tunneling microscopy (STM)** or by complicated and sophisticated quantum mechanical calculations.

It is interesting to mention that there is a correlation between the molecular structures of alkanes and some of their physical properties. By correlating the number of carbon atoms in simple alkanes with the melting points of the same compounds we observe that the molecules with odd numbers of C-atoms and those with even numbers of C-atoms exhibit different correlation curves.

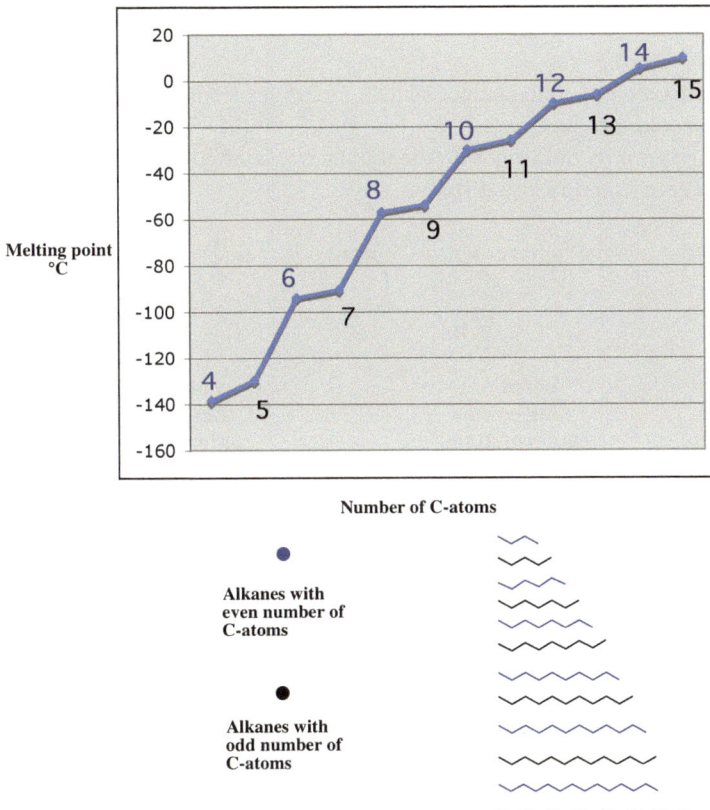

Number of C-atoms

Alkanes with
even number of
C-atoms

Alkanes with
odd number of
C-atoms

Looking at the structures of alkanes in the figure we can see that while the molecules with an odd number of carbon atoms appear symmetrical (both terminal carbon-carbon bonds being oriented upwards - black lines), where the structures with an even number are asymmetrical in the sense that the terminal carbon-carbon bonds are oriented upwards on the left end of the chain, but downwards on the right end (blue lines). Although the correct explanation is not simple, this example demonstrates how macroscopic properties correlate with microscopic structures. We can assume that "odd" molecules in the condensed state shall be packed differently from "even" molecules.

1.3   Molecular Dynamics and Conformations

The consideration of long-chained alkane molecules leads to the question of whether these molecules are flexible? To analyze this flexibility, let us take the ethane molecule as an example. Starting with the tetrahedral configuration, the ethane molecule could be represented in two different ways:

Form A
Staggered
conformation

60°

Form B
Eclipsed
conformation

The form A can be transformed into B simply by rotation of one of the methyl groups by the torsion angle $\Phi=60°$ around the C-C bond. Such rotations, especially at room temperature, are very fast and it is possible to imagine that the ethane molecule could appear in a large number of shapes, depending on the rotation angle Φ. Such different shapes of molecules which follow from internal rotations about single bonds are called **conformations**. In the figure above the two most important conformations: **staggered** and **eclipsed** are shown using the wedge-dash notation. However, these conformations can also be drawn by rotating the molecule by 90° relative to the plane of drawing. In that case the carbon atoms appear one behind the other. As shown in the following figure, the carbon atoms in the front and at the back are separated by a circle. The representation which is shown in the next scheme is called a **Newman formula**.

Newman formulas serve to clarify the difference between staggered and eclipsed conformations. In the staggered conformation, the C-H bonds at the neighboring carbon atoms are closer to each other than in the eclipsed conformation. Since the electron clouds in these covalent bonds are negatively charged they lead to a repulsive interactions so that the eclipsed conformation has a higher potential energy than the staggered conformation. We can say that the neighboring C-H bonds **sterically hinder** each other, which gives the molecule in the eclipsed conformation a higher potential energy.

Staggered
conformation

60°

Eclipsed
conformation

NEWMAN FORMULA

Starting with the eclipsed conformation, if the torsion angle increases the potential energy decreases up to the angle of 60°, which is characteristic for the staggered conformation. The dependence of the potential energy on the torsional angle is shown in the following scheme.

We can observe that the potential energy varies periodically with the angle and that the staggered conformations always correspond to the energy minima while the eclipsed conformations always corresponding to the energy maxima. It is for this reason that ethane molecules assume the staggered rather than the eclipsed conformation.

Stable conformations which correspond to the energy minima are called **conformers**. If the molecule comprises more than one C-C bond the rotation is possible around all of them and the number of conformers increases. Let us investigate the conformations and conformers of the butane molecule:

By investigating the conformations of butane and considering only the rotation around the central C-C bond we can see that three conformers are possible (labeled **B, D** and **F** in the scheme above). The most stable conformer is **D** because in this case the bulky methyl groups, which have dense electron clouds, are at the largest distance from each other, which makes their **steric hindrance** smallest. Conformer **D** has the smallest potential energy and we can say that it corresponds to the **global minimum** of the potential energy curve. Less stable conformers **B** and **F** correspond to **local minima** of the potential energy curve. Transformation of one conformer into another is possible only by passing through the potential energy maxima **A, C, E** and **G**, which belong to eclipsed conformations. As the molecule gets more complex, the number of conformers (local energy minima), as well as the number of local maxima (eclipsed conformations) becomes larger. Exploring the shapes and energies of possible conformers and the energies of eclipsed conformations requires special calculations called **conformational analysis**. The simplest and most often used method is based on the mechanical model in which the atoms are defined as mass points and the chemical bonds are treated as elastic connectors. Using this model it is possible to calculate the energy minima and maxima. This method is called **molecular mechanics**.

1.4   Cycloalkanes

The idea that carbon atoms can be bound in cyclic structures (besides forming chains and branched chains) appeared during the second half of nineteenth century. When these cyclic structures are saturated hydrocarbons we talk about **cycloalkanes**. In the nomenclature of this class of compounds the name is formed by adding the prefix **cyclo-**.

Cyclopropane   **Cyclopentane**   **Cyclooctane**

Configuration of cycloalkanes is based on the combinations of tetrahedrons, similarly to the case of alkanes. However, in some cases the ring structure requires that the angles between C-C bonds deviate from the normal tetrahedral values (109°28′). Almost ideal tetrahedral angles are present in the molecule of cyclohexane. For the pictorial representation of three dimensional molecular structures of cyclic molecules special descriptive projection is used. Let us analyse the structure of cyclohexane molecule in more detail.

1.4.1 Cyclohexane

As we can see in the following scheme, the hydrogen atoms can be bound to carbons in two different ways. When C-H bonds (containing green H atoms) are parallel to the axis perpendicular to the plane of the ring, hydrogens in such bonds are called **axial** hydrogens (shown in green). Hydrogens drawn in red lie nearly in the equatorial plane of the ring and are called **equatorial** hydrogen atoms. The difference between these types of hydrogens is better represented by using the Newman formula:

The Newman formula shown in the upper scheme (right) is obtained by looking along the C_3-C_4 and C_1-C_6 bonds. On inspection it becomes clear that the represented conformation has a staggered form and that these are stable conformers. By convention they are called **chair conformers**. Since all the C-C bonds are in the staggered arrangements the cyclohexane molecule is the most stable cyclic hydrocarbon. There are no C-H bonds that sterically hinder each other. Conformational dynamics of the cyclohexane ring involves limited rotation around the C-C bonds: one chair conformer is transformed into another.

Hydrogen atoms (blue) that are in the axial positions in one conformer appear in equatorial positions in the product conformer. Such change from axial to the equatorial position is even more evident in the methyl-substituted cyclohexane.

More stable chair conformation

The methyl group is in the axial position in the first conformer and in equatorial position in the second conformer. Both conformers can also be represented by Newman formulas:

More stable conformer

In the left conformer (scheme above) the axial methyl group is close to the nearby axial C-H bond. Since the electron clouds of these two axial groups repel each other, the molecule posses high energy and becomes less stable. As we have already commented, such unfavorable interaction is called steric hindrance. In the conformer on the right, the methyl group is in equatorial position, the steric hindrance does not exists and the conformer is more stable. This is the explanation of the general rule according to which the conformers of cyclohexane with large substituents in the equatorial positions are more stable than the conformers with groups in the axial positions.

1.4.2   Cyclopentane

In contrast to cyclohexane the cyclopentane molecule is almost planar. While four carbon atoms lie in the molecular plane the fifth atom is slightly distorted out of the plane. Conformational dynamics of the cyclopentane ring is almost negligible and involves sequential out-of-plane lifting of C-atoms. The molecule in the following figure is represented in two ways, the accurate description on the left hand side and a simplified on the right.

Simplified representation

Since all the hydrogen atoms are quasi-axial the representation of cyclopentane molecule by a planar ring is an acceptable description.

1.4.3 Cyclobutane and Cyclopropane

While the angles around carbon atoms in cyclohexane and cyclopentane do not deviate much from the tetrahedral angle in small ring molecules such as cyclobutane or cyclopropane the geometry requires that the angles between neighboring C-C bonds deviate significantly from the tetrahedral value. Such forced reduction of tetrahedral angles requires additional potential energy. Consequently, small cyclic molecules such as cyclopropane contain an excess potential **strain energy** called the **Bayer strain**. Because of this strain, molecules with three and four membered rings are chemically active and in chemical reactions the rings tends to cleave and relieve excess potential energy.

| Cyclobutane | Cyclopropane |

The cyclobutane molecule is not planar and the carbon atoms deviate from the plane by forming a structure in which four of the hydrogen atoms have an equatorial position (labeled in red) and the remaining four hydrogens are axial (labeled in green). In this way the Bayer strain is minimized. In cyclopropane, all the hydrogens are in quasi-axial positions.

From the detailed studies of the cyclopropane molecule, it has been found that the C-C bonds are folded i.e. the electron density maxima between bonded carbon atoms lie off the C-C bond axis.

1.5 Polycyclic Hydrocarbons

Carbon atoms can also be interconnected in structures that have more than one ring. They are called **polycyclic hydrocarbons**. One well known polycyclic structure that appears in a series of natural products is **decaline**, with two condensed six membered rings. The term "condensed" means that rings share one C-C bond. Since there are two different hydrogen positions on this common C-C bond, the decaline molecule can appear as two stereoisomers. The neighboring hydrogens can be either on the same side or on the opposite sides of the shared C-C bond and the stereoisomers are known as *cis*-**decaline** and *trans*-**decaline**, respectively. As expected, the isomers have different physical properties. The melting point of *cis*-decaline is 198 °C and the melting point of *trans*-decaline is 185 °C.

Decaline   *cis*-**Decaline**   *trans*-**Decaline**

In chemical nomenclature, the polycyclic hydrocarbons are named regarding the size of the bridges. The carbon atoms which belong to two different rings are called **bridgeheads** but they do not form part of the bridge. The suffix **bicyclo-** is used if there are two rings and the number of C-atoms on the bridges is added in parenthesis.

Bridgehead carbon atom

Bridgehead carbon atom

Bicyclo[2.2.2]octane   **Bicyclo[4.3.2]undecane**

Let us describe some frequently encountered polycyclic structures. **Bicyclo[2.2.1] heptane**, also known as **norbornane**, is the basic structure of a series of natural products. The tricyclic compound **adamantane** (tricyclo[3.3.1.1^{3,7}]decane) is in principle the smallest structural fragment of diamond and its structure appears on the list of pharmaceutically important products. Adamantane was synthesized for the first time in 1941 at the University of Zagreb by **Vladimir Prelog**.

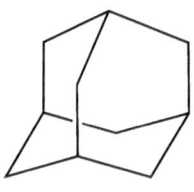

Bicyclo[2.2.1]heptane
(Norbornane)   **Adamantane**

Chapter 2
Functional Groups

As we have seen in the previous chapter, the hydrocarbon skeleton is responsible for the shape and flexibility of organic molecules. In the case of alkane molecules, the molecular structure is based on tetrahedral units and the molecular dynamics is the consequence of relatively free rotations about the carbon-carbon single bonds. These rotations give rise to different conformations. However, with the exception of small-ring molecules, the alkanes, as compounds containing only carbon and hydrogen, are relatively weakly reactive substances.

Most organic molecules which exhibit chemical reactivity have an incorporated active structural unit called the **functional group**. In the structural formula the unspecified group, **the substituent**, bound to the hydrocarbon skeleton is labeled as R. To be chemically active, the functional group must possess high energy electrons which can be either the electrons in the multiple bonds or the non-bonded electrons on atoms other than carbon or hydrogen. Such atoms (for instance O, N, S, P, Cl, Br, I *etc.*) when present in the organic molecules are called **heteroatoms**. The presence of the functional group is also the basis for systematization of organic compounds into specific classes.

The most common functional groups together with their nomenclature are listed in following tables. The additional functional groups as well as the details of the nomenclature of specific classes of compounds will be discussed later in this book.

H. Vančik, *Basic Organic Chemistry for the Life Sciences*,
DOI 10.1007/978-3-319-07605-8_2, © Springer International Publishing Switzerland 2014

Amongst the functional groups which have double and triple bonds we shall mention those in the following table. All the compounds belong to different types of hydrocarbons:

ALKENES	-ENE		PROPENE
ALKYNES	-INE		PROPYNE
AROMATIC COMPOUNDS			BENZENE

The main functional groups and the corresponding organic compounds containing oxygen are: alcohols, ethers, ketones, aldehydes, carboxylic acids and esters as listed in the following table:

ALCOHOLS	-OL	CH_3CH_2OH	ETHANOL
ETHERS	ETHER	$CH_3CH_2OCH_2CH_3$	DIEHYL ETHER
ALDEHYDES	-AL		ETHANAL
KETONES	-ONE		PROPANONE
CARBOXYLIC ACIDS	-OIC ACID		ETHANOIC ACID
ESTERS	-YL -OATE		METHYL ETHANOATE

The most important organic compounds with nitrogen are the following:

AMINES	-AMINE AMINO-	$CH_3CH_2NH_2$	ETHYL AMINE AMINOETHANE
NITRILES	-NITRILE	CH_3CN	(ETHANONITRILE) ACETONITRILE
AZO COMPOUNDS	AZO-	$H_3C-N\!\!=\!\!N-CH_3$	AZOMETHANE
DIAZO COMPOUNDS	DIAZO-	$H_2C\!\!=\!\!N^+\!\!=\!\!N^-$	DIAZOMETHANE

Some organic compounds possess groups with both elements, oxygen and nitrogen:

AMIDES	-AMIDE		ETHANAMIDE
NITRO COMPOUNDS	NITRO-		NITROMETHANE
NITROSO COMPOUNDS	NITROSO-		NITROSOMETHANE

Let us also list the classes of compounds with the functional groups containing sulfur and halogen:

HALIDES	**-IDE** **HALO-**	CH_3Cl	**METYL CHLORIDE** **CHLORMETHANE**
THIOLS	**-THIOL**	CH_3CH_2SH	**ETHANTHIOL**
SULFIDES	**-SULFIDE**	$H_3C\overset{S}{\diagdown}CH_3$	**DIMETHYL** **SULFIDE**
SULFOXIDES	**-SULFOXIDE**	$H_3C\underset{S}{\overset{O}{\diagdown}}CH_3$	**DIMETHYL** **SULFOXIDE**
SULFONES	**-SULFONE**	$H_3C\underset{S}{\overset{O\ \ O}{\diagdown}}CH_3$	**DIMEHYL** **SULFONE**

Chapter 3
Electronic Structure of Organic Molecules

3.1 The Covalent Bond

More than a 100 years ago, **Joseph John Thomson** who in 1897 discovered the electron, hypothesized that this small negatively charged particle plays crucial role in the formation of the chemical bond. Thomson argued that atoms in the chemical bond exchange electrons in such a way that every atom donates one electron to the common chemical bond. Consequently, the electron pair is responsible for holding the two atoms together. However, the proposed model was not able to explain why the bonds in molecules containing the same atoms such as H_2 are different from the bonds in molecules with different atoms, for example HCl or NaCl. The idea about covalent and ionic bonds appeared in 1916 when **Gilbert Newton Lewis** proposed the representation of the chemical bond as the common electron pair shared by the two bound atoms. The electrons in Lewis pairs do not have any identity regarding the atoms from which they originate and the covalent bond can be represented by dots as follows:

$$\text{H} \cdot \qquad \cdot \text{H} \qquad \text{H} : \text{H}$$

Using this model the methane molecule is represented as:

$$
\begin{array}{c}
\text{H} \\
.. \\
\text{H} : \text{C} : \text{H} \\
.. \\
\text{H}
\end{array}
$$

Traditionally used valence lines acquire new meaning: they represent electron pairs. Within this semantics, the double bond is labeled by two lines which describe two electron pairs. Commonly, such notation is called the **Lewis structure**.

H. Vančik, *Basic Organic Chemistry for the Life Sciences*,
DOI 10.1007/978-3-319-07605-8_3, © Springer International Publishing Switzerland 2014

$$\ce{H2C=CH2}$$

However, in some molecules such as ammonia, some of the electrons are not included in the chemical bonds. Out of five valence electrons on the nitrogen atom, only three take part in covalent bonds with hydrogens, the remaining two are **nonbonding** and are called the **electron lone pairs**.

$$\ce{H-\underset{H}{\overset{..}{N}}-H}$$

Molecules with electron lone pairs are very reactive. Since organic chemistry is based on carbon compounds and carbon atoms only have electron lone pairs in special cases, the carriers of nonbonding electrons are heteroatoms such as N, O, S, P or halogens.

$$\ce{CH3CH2\overset{..}{\underset{..}{O}}H} \qquad \ce{CH3CH2CH2\overset{..}{N}H2} \qquad \ce{CH3-\underset{CH3}{\overset{CH3}{C}}-\overset{..}{\underset{..}{Cl}}{:}}$$

The examples of carbon atoms with electron lone pairs are the cyanide ion: CN⁻ or the relatively unstable but reactive **carbenes**, which are common intermediates in photochemical reactions.

Although the Lewis model has been accepted as a basic and universal concept for the description of the constitutions of molecules, this representation appears to be inadequate for some species. For instance, by using the Lewis model for the description of the nitrate anion NO_3^-, as shown in the following figure, one of the oxygen atoms is bound to the nitrogen via a double bond and the two remaining oxygens via a single bonds. Since it is known that double bonds are shorter than single bonds (this will be discussed later in the book) the proposed description of the nitrate ion indicates that one of the nitrogen-oxygen bonds should be shorter than the two remaining bonds. However, since precise measurements show that all three nitrogen-oxygen bonds are of equal length, the constitution of the nitrate ion cannot possibly be explained by a single Lewis formula. At the beginning of the twentieth century **Sir Robert Robinson** and **Fritz Arndt** have proposed that such molecules can be interpreted with a set of structural formulas, which have different electron configurations. This particular electron configuration is called the **resonance structure** and we could say that the molecule is better represented by several resonance structures. In literature we can find the statement that the molecule is a **resonance hybrid** of the corresponding canonic resonance structures. In our example, the nitrate ion is better represented by three canonic resonance structures.

By definition, all resonance structures are descriptions of the same molecule. We use the notation in which the different resonance structures are connected by double tipped arrows with all the formulas placed within parentheses.

As shall be demonstrated later in this book, the knowledge of resonance structures can help with the prediction of important molecular properties such as charge distribution or the nature of a particular bond. Canonical resonance formulas can be constructed by using special rules:

1. Since the canonic structures are descriptions of the same molecule, the atom positions in all the resonance formulas must be unchanged.
2. The total charge must also be unchanged. For instance, in the nitrate ion the total charge in all the structures is −1.
3. Only the electrons and electron pairs can be shifted. In principle, the electron lone pair in one resonance structure becomes the bonding pair in another structure and *vice versa*:

Robinson arrow

By convention, the shift of electron pairs is indicated by special arrows called the **Robinson arrow**, in the honor of **Robert Robinson**, the chemist who developed the theory of chemical reaction mechanisms.

One further conclusion can be drawn from the resonance model. The canonical resonance forms suggest that electrons do not have fixed positions within the molecule; sometimes they appear as lone pairs, but sometimes they comprise double bonds. This demonstrates that electrons are distributed throughout the entire molecule or at least throughout its major part; the electrons are **delocalized**. It is important to point out that the **delocalization** of electrons decreases their energy so that such delocalized electrons have a lower energy than the electrons which are localized, for instance in the single bond or on the heteroatom (lone pairs on nitrogen or oxygen). In this way, by constructing the canonic resonance formulas we are able to compare the stability of different molecules.

3.2   Molecular Orbitals

Although the Lewis concept together with the resonance model forms the basis for unambiguous descriptions of structures of molecular classes and for organic reactions, their use in understanding and explaining the nature of the chemical bond is insufficient. For a detailed description of the nature of electrons we need the theory that describes electrons as waves. Such theory is called **quantum mechanics** and was created in 1925 by **Werner Heissenberg** and in 1926 by **Erwin Schrödinger**. In quantum mechanics, the electrons are entities which could behave as either particles or waves, depending on the type of the experiment by which we observe them. We say that the electrons have a **dual nature**.

For the study of electrons in atoms and molecules it is more convenient to consider electrons as waves. Since the electrons are contained within atoms or molecules, the waves that describe them are **standing waves**, as are for instance the waves of water which move within a closed pool. The standing wave also describes the vibration of a strained wire for example on some musical instrument. Let us use the analogy of a strained wire to obtain a deeper insight into the behavior of electrons in molecules.

After triggering, the wire vibrates with a certain frequency and we hear the sound of the corresponding note. If the wire is shorter the note (frequency) is higher and vice versa, the longer the wire the lower the note (frequency). As we know from physics, the vibrational frequency is proportional to the energy and consequently, the shorter wire has higher energy than the longer one. Taking this rule into account, by connecting two shorter wires to construct one longer wire we shall get a lower energy. We conclude that the energy is lower if the standing wave moves over a larger area of space. This is described in the following diagram:

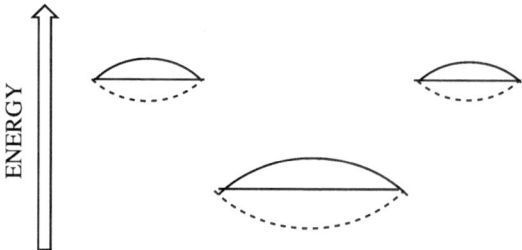

If the wave function is labeled with Φ, than the lower energy wave is obtained by adding the functions Φ_1 and Φ_2. Imagine that the wires described by Φ_1 and Φ_2 merge into a single wire which has twice the length and that we press the resulting longer wire in the middle (as we do for instance when playing a string instrument). The resulting wire will then vibrate with the opposite phases, left and right from the pressure point. Such a system could be represented as the difference of the functions Φ_1 and Φ_2. The result is two composite waves different in phase but close to each other in energy. Let us represent them by $\Phi_1 - \Phi_2$. If the smaller wire-waves are both electrically charged with the same charge (for instance negative), their common energy should be higher since the negative charges repel each other.

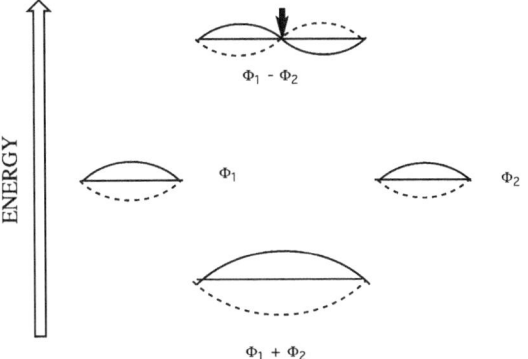

In principle, the two resulting standing waves can be obtained from Φ wave functions in two ways: in phase combination ($\Phi_1 + \Phi_2$) with lower energy and out of phase combination ($\Phi_1 - \Phi_2$) with higher energy. In atoms the standing waves, which describe electrons, are represented by special functions called **atomic orbitals**. Here we will not discuss the details of such mathematical functions. Rather we will use their graphic representations. Since carbon and hydrogen are the most important atoms in organic chemistry, we will represent their atomic orbitals only. While the electron in the hydrogen atom is present in only one orbital (1s), the electrons in the carbon atom are distributed in **1s, 2s, 2p$_x$, 2p$_y$ and 2p$_z$ orbitals**. From the following figure is clear that the $2p_x$, $2p_y$ and $2p_z$ orbitals are of equal energy, which is however higher than the 2s orbital energy.

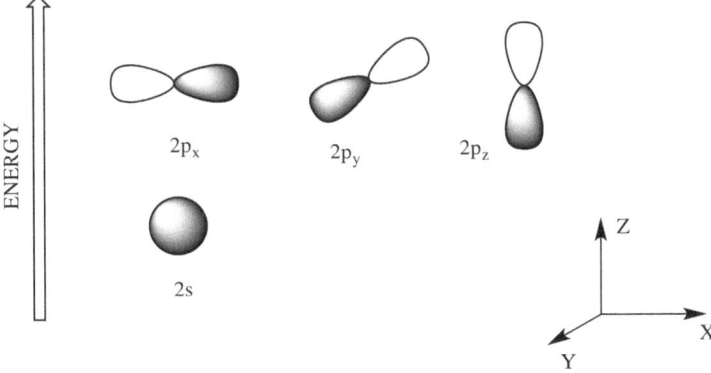

The p-orbitals ($2p_x$, $2p_y$ and $2p_z$) differ only by their orientation in space along the coordinate axes x, y and z. The different colors of the two sides of each p-orbital (gray or white) corresponds to the different wave phases, similar to as was discussed previously for the wire model.

Next we have to combine the atomic orbitals to construct the molecule in the same way as we have done with wires representing standing waves. In the simplest case, the formation of the hydrogen molecule, we have two combinations of atomic orbitals: the lower energy "1s + 1s" combination and the higher energy "1s − 1s" combination. These two combinations of atomic orbitals describe electrons in the molecule and they are called **molecular orbitals**. In analogy with our standing wave model, these combinations, i.e. the molecular orbitals, have different phase relations: (bonding and antibonding). The appearance of electrons in the antibonding orbital converts the helium molecule back into its corresponding atoms. For this reason the noble gases do not form molecules.

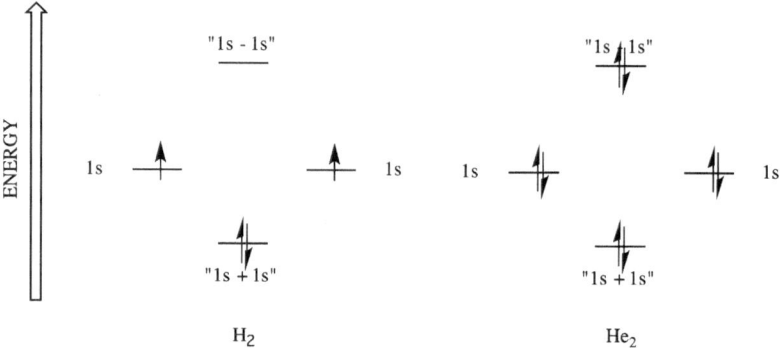

Bonding between carbon atoms could be described analogously, except that $2p_x$ orbitals are used instead of 1s orbitals. Combining $2p_x$ orbitals, we get bonding ("$2p_x + 2p_x$") and antibonding ("$2p_x − 2p_x$") molecular orbitals:

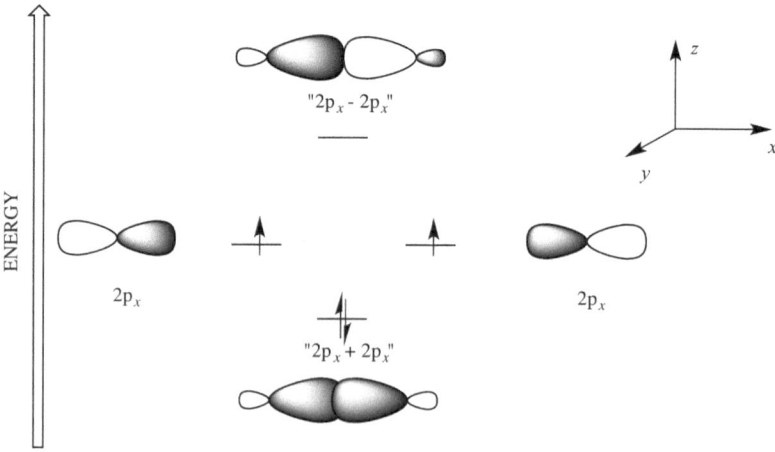

Since 2p orbitals can have three different spatial orientations along three coordinate axes, it is necessary to consider the other two combinations: "$2p_y \pm 2p_y$" and "$2p_z \pm 2p_z$" as well. For convenience let us discuss the "$2p_z \pm 2p_z$" case first.

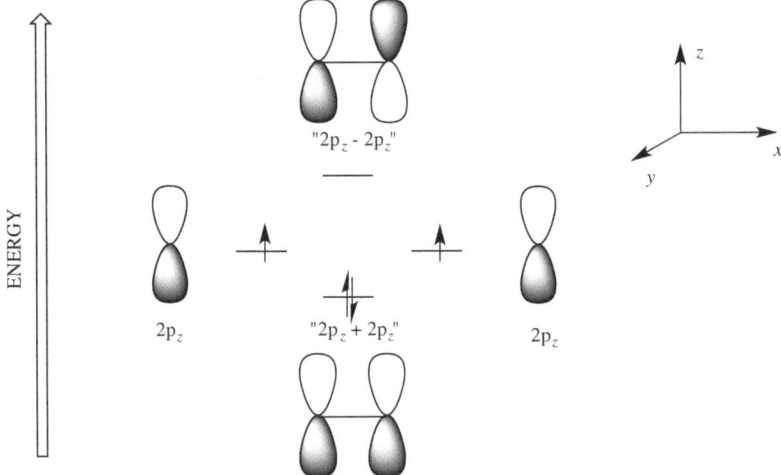

We obtain two molecular orbitals: bonding "$2p_z + 2p_z$" and antibonding "$2p_z - 2p_z$". The same reasoning would also give pairs of "$2p_x \pm 2p_x$" and "$2p_y \pm 2p_y$" orbitals. What is the main difference between ("$2p_z \pm 2p_z$") or ("$2p_y \pm 2p_y$") and ("$2p_x \pm 2p_x$") orbitals. By rotating individual atomic orbitals in the "$2p_x \pm 2p_x$" combination around the bond axis, the properties of the bond remain the same. However, by performing the same rotation on the "$2p_z \pm 2p_z$" combination the phases (depicted with gray and white shading) are exchanged as demonstrated in the next figure.

Using this property, the orbitals can be classified into two types: orbitals which are symmetric upon the rotation around the bond-axis are called **σ-orbitals** and orbitals which are antisymmetric are called **π-orbitals**.

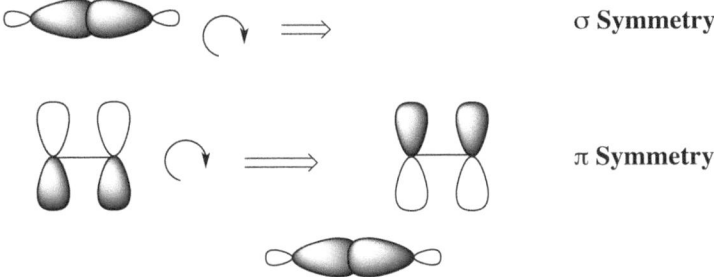

σ Symmetry

π Symmetry

The same reasoning applies to the antibonding combinations. The antibonding orbitals are labeled with the symbols σ* and π*.

The extent of the stabilization of bonding orbitals or the destabilization of antibonding orbitals, depends on the degree of orbital **overlap**. In σ-orbitals the overlap between $2p_x$ and $2p_x$ atomic orbitals is greater than the overlap between

$2p_z$ and $2p_z$ in π-orbitals. For this reason the σ -orbital has lower energy than the π-orbital and the antibonding σ* has higher energy than the π*.

This knowledge can help us in understanding the electron distribution in organic molecules. The lowest in energy are σ-orbitals, followed by π-orbitals, nonbonding **n**-electrons (the electron lone pairs), π* orbitals and the highest energy σ* orbitals. In the electronic ground states σ, π and n orbitals are occupied by electrons while π* and σ* orbitals are vacant. The nonbonding **n**-orbital in principle does not contribute to the chemical bond. All the molecular orbitals are shown in the following figure. However, two of them have the additional labels HOMO, and LUMO. These two orbitals are called **frontier orbitals**, because they are especially important for analyzing possible chemical reactions. The HOMO acronym stands for the **h**ighest **o**ccupied **m**olecular **o**rbital and LUMO for the **l**owest **u**noccupied **m**olecular **o**rbital.

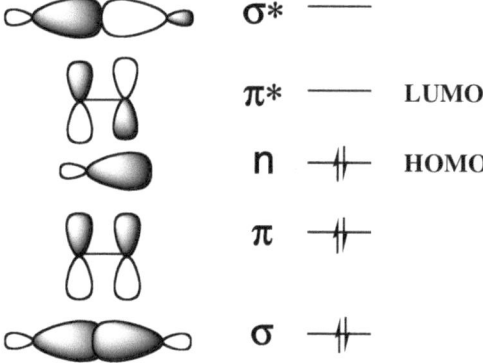

The molecule is in the **ground electronic state** if all valence electrons occupy only bonding or n-orbitals. If the molecule absorbs a quantum of visible or ultraviolet light and if the energy of this quantum corresponds to the energy difference between the bonding and the antibonding orbitals (for instance between π and π*) the electrons can be moved from the bonding to the antibonding orbital and the molecule shall go to the **excited electronic state**. Such excitation is the main characteristic of the reactions induced by light, the photoreactions.

3.3   Distribution of Electron Density, and the Shape of Molecules

Description of the electronic structure of molecules by using this molecular orbital model provides important information about the nature of molecules. One of the most important pieces of information is the electron density distribution. While in σ -bonds the electron density is largest between the atoms i.e. "along the line of the chemical bond" in π bonds the electron density is concentrated not between the atoms but above and below the plane which contains the "bond line". This is represented in the next figure. In an ethane molecule, which possess only σ -bonds, the electron cloud is situated between the C-atoms. However, in the ethene molecule, with a π-bond, the electron density is highest above and below the plane in which all the atoms are located. Electron clouds in ethene are not blocked by the atoms and are situated on the open side of the molecule where other particles can attack it. Consequently an ethene molecule is much more chemically reactive then ethane.

ETHANE   ETHENE

Although a lot of information about molecular properties can be obtained from the molecular orbital model, for deeper insight into the electronic structure of molecules we need a much more sophisticated approach. The methods include quantum-mechanical (or to use a better term quantum-chemical) calculations based on fundamental mathematical and physical concepts. Molecular shape, electron density and charge distribution calculated at such levels provide a much more realistic picture of the molecules. In the following figure on the left hand side the methane molecule is presented by using the simple ball-and-stick model, while on the right hand side the calculated shape of the same molecule is shown. The green surface represents the surface of the maximum electron density.

3.4   Bond Lengths, Bond Energy, and Molecular Vibrations

The length of the chemical bond depends primarily on the atomic radii of the atoms included in the bond and on the electron interactions. Since in organic chemistry we deal mostly with carbon, hydrogen and a few heteroatoms we will focus predominantly on the bond-lengths between these atoms. As is evident from the table later in this chapter single, bonds are in principle longer than double bonds, which are in turn longer than triple bonds. Additionally, multiple bonds are stronger than single bonds. However, interatomic distances in chemical bonds are not fixed as implied by the rigid stick-and-ball models. A much better representation would be a mechanical model in which the balls are interconnected by springs. In such a model the massive balls are not fixed but are free to vibrate around the equilibrium position. Analogously, the atoms in the molecule vibrate around the equilibrium point at a certain frequency. The chemical bond-length is defined as the interatomic distance when the atoms are in the equilibrium position. Vibrational frequencies of atoms in chemical bonds are nearly hundred times smaller than the frequencies of electrons-waves. Consequently, in order to become vibrationally excited, the molecule must absorb electromagnetic radiation of longer wavelengths than visible light, which is the radiation in the infrared region.

The simplest molecular vibrations are bond stretchings. Besides stretching, there are other vibrational modes, for instance vibrations of changing the angle between bonds. Let us analyze the vibrations of a molecule of water. As shown in the next scheme, there are two vibrational modes for bond stretching v_1 and v_2 and one vibrational mode for the angle change v_3.

Vibrations v_1 and v_2 differ in symmetry. While for v_1 we can say that it is symmetric because both OH bonds stretch simultaneously, v_2 is antisymmetric since the H-atoms move in opposite directions. Hence, the water molecule possesses three vibrational modes. Generally, if the molecule has N atoms it will have 3 N-6 vibrational modes.

Functional group	Wave number in cm^{-1}	Mode of vibration
CH$_3$, CH$_2$, CH, alkanes	2,830–2,990	C-H stretch
=CH$_2$, CH alkenes	3,000–3,100	C-H stretch
CH$_3$, CH$_2$, CH, alkanes	1,300–1,450	HCH bending

(continued)

Functional group	Wave number in cm^{-1}	Mode of vibration
C=C, alkenes	1,600–1,660	C=C stretch
C≡C, alkynes	2,000–2,150	C≡C stretch
C=O, aldehydes, ketones, carboxylic acids, esters	1,700–1,780	C=O stretch
C-O, alcohols, ether	1,150–1,280	C-O stretch
C-Cl, alkyl-chlorides	420–700	C-Cl stretch

Each vibrational mode has its characteristic frequency and the molecule can absorb infrared light of frequencies which correspond to the vibrations that are active in the molecule. This fact is used for structural characterization of molecules. Frequencies of the absorption maxima in the infrared spectrum allow us to identify the vibrational modes. Since every functional group possesses characteristic vibrations, the infrared spectrum can be used to identify the functional groups present in the molecule. This method is called **infrared spectroscopy**. In spectroscopy the vibrational frequencies are represented as wavenumbers which are measured in units called reciprocal centimeters cm^{-1}. The reciprocal centimeter unit is the number of waves within 1 cm. Some characteristic vibrational wavenumbers for the most important functional groups are presented in the table above.

In the figure below the infrared spectrum of acetone, CH_3COCH_3 is shown

Signals at 2,966 cm^{-1} are assigned to the CH stretching of the methyl groups, the sharp maximum at 1,749 cm^{-1} is the characteristic stretching of the CO group and peaks at 1,363 cm^{-1} correspond to the change of the angle between the CH bonds in the methyl groups (bending).

Assignment of all the peaks is difficult because the vibrations of the acetone molecule are complex: the molecule has 10 atoms and by using the formula 3N-6 it must have $3 \times 10 - 6 = 24$ vibrational modes.

Now, we must return to the simplest vibration, the stretching of the bond in a diatomic molecule e.g. in HCl. Imagine that the H-Cl bond is stretched continuously in one direction. From the mechanical point of view the stretching of the spring increases the potential energy. At some distance the bond will break. Potential energy increases also by compressing the bond. Relationship between the potential energy and bond length is represented in the next figure. Such curves are known as **Morse curves**. As shown in the following figure at interatomic distances larger than a certain value, the potential energy becomes constant because the chemical bond is broken. The distance at which the potential energy has minimum value is associated with the length of the chemical bond. In reality, the molecule is never at this minimum because the molecule constantly vibrates and the energy of this vibration must be larger than this minimum. The difference between the minimum energy and the vibrational energy level is called **zero point energy** and it depends on the frequency of vibration by the formula $E = \frac{1}{2}(h\nu)$. h is **Planck constant** and ν is the vibrational frequency. If the molecule is cooled down to the temperature of absolute zero than all its vibrations will have energy $\frac{1}{2}(h\nu)$.

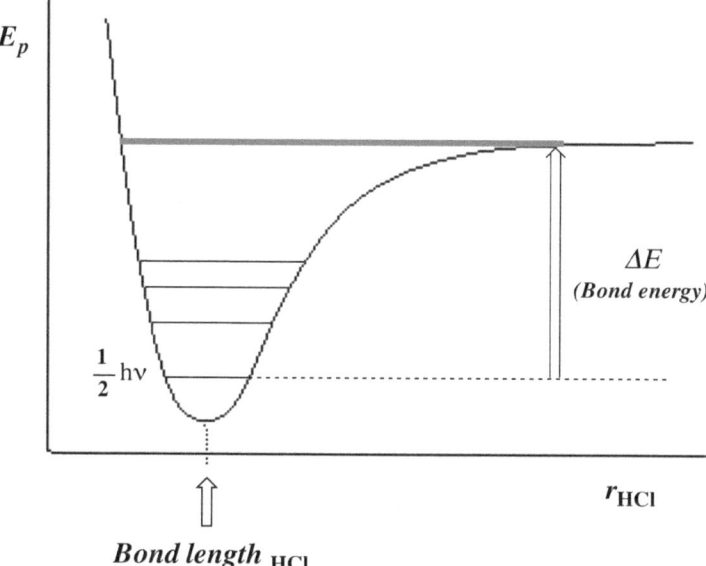

E_p

$\frac{1}{2} h\nu$

ΔE
(*Bond energy*)

r_{HCl}

Bond length $_{HCl}$

In analogy with the energy of electrons, the energy of molecular vibration cannot have any value, but is distributed amongst the accessible quantum energy levels. Such energy levels have defined values and the distance between them diminishes as the potential energy increases. If the energy reaches ΔE the potential energy becomes continuous (see the previous figure) and the chemical bond is broken. Hence, the energy ΔE required to break the chemical bond is called the

dissociation energy. There is close relation between the dissociation energy and the **bond energy** i.e. the energy necessary to break the chemical bond. Strength and energy of the chemical bond depends on the energy difference between the zero point energy and the energy of continuum. If a chemical bond is strong, the potential energy curve is steep and the bond energy (related to ΔE) is larger, so the position of the energy minimum is at the shorter bond distance. Conversely, weak bonds have wider potential energy curves, a small ΔE and the positions of their minima are at large interatomic distances.

Typical values of bond energies and bond distances for the chemical bonds in organic molecules are listed in the following table. Please note that the values in the table are averages, based on the measurements for a large number of molecules. Accurate parameters for individual molecules can be obtained from high level quantum chemical calculations or by the use of sophisticated instruments.

Covalent bond	Bond length in pm	Bond energy in kJ/mol
C-H	106	413
C-C	154	347
C=C	134	614
C≡C	121	811
C-O	113	351
C=O	116	711
C-Cl	163	326
C-Br	191	276
O-H	94	464
N-H	98	389
S-H	132	339
Cl-H	127	431

The important conclusion from this discussion is the rule that the bond energy and bond length are in an inverse relationship: as the bond gets shorter it becomes stronger and *vice versa*, as the bond is weaker its energy is smaller. This is important information, because from the data about bond lengths it is possible to predict which bond is weaker and consequently which is more prone to breaking in a chemical reaction.

In principle, by using values such as those in the table above, it is possible to predict molecular geometry and from this the possible reactivity of molecules. Additionally, the sum of all bond energies in the molecule represents approximate value of the **enthalpy of formation** of the corresponding compound. For example in a molecule of ethanol CH_3CH_2OH we count five C-H bonds, one C-C bond, one O-H bond and one C-O bond, so the approximate enthalpy of formation, ΔH_f, is therefore: $5 \times 413 + 347 + 464 + 351 = 3{,}227$ kJ/mol.

3.5 Deducing Molecular Structure by Nuclear Magnetic Resonance Spectroscopy

Besides infrared spectroscopy the most convenient method for deducing the molecular structure is **nuclear magnetic resonance spectroscopy** (**nmr**). The basic principle of this method is the detection of changes in the orientation of the atomic nuclear spin of hydrogen (the spin of a proton) and carbon (the spin of the nucleus of the ^{13}C isotope).

Like an electron, the proton also has an angular momentum (**spin**) which is something akin to the rotation around its proper axis. Since the proton is a charged particle, its rotation induces a magnetic field. Roughly speaking, the proton resembles a small magnet. Let us imagine the model in which we have two such small magnets with different orientations. If the magnets are distant from each other their energies are equal, independent of their mutual orientation. However, inside an external magnetic field the proton whose orientation is favored relative to the external field has a lower energy and the proton with the opposite orientation has a higher energy. This energy difference is shown on the figure below as ΔE.

The higher energy proton can change its orientation from disfavored to favored by radiating a quantum of energy. This radiation has a corresponding frequency ν:

This emitted radiation falls within the radio wave or microwave frequency region. By using a radio wave receiver which can detect this radiation we can detect this quantum transition. However, the transition is so fast that at the moment when the sample is placed into the magnetic field all the protons will reorient themselves so quickly that is becomes very difficult to register signals. To avoid this difficulty the instrument, the nmr spectrometer has an oscillator which irradiates the sample with radio or microwaves by which the spins are constantly reoriented into disfavored orientations. The basic principle of the construction of nmr spectrometer is represented in the next figure.

The frequency or radiation depends on the energy difference between nuclei H_A and H_B (middle of the figure above). However, the protons (nuclei of hydrogen atoms) exist within molecules and are surrounded by electron clouds of different densities. Such electron density clouds can reduce the influence of the external magnetic field on the energy required for the reorientation of proton spins. We can say that the electrons **shield** the atomic nucleus. The extent of this shielding depends on the functional group to which the hydrogen atom belongs. Thus the hydrogen in a CH_3 group is more shielded than the hydrogen in the CH_2 group. Hydrogen in the $=CH_2$ is shielded even less:

Consequently in the nmr spectrum of the molecule which contains all three groups (CH_3, CH_2, and $=CH_2$) three signals at ν_1, ν_2 and ν_3, which can be assigned to the three different H-atoms, will appear. This spectroscopy detects hydrogen atoms in different environments in the molecule and is called the [1]**Hnmr spectroscopy**.

In most nmr spectra the units which are related to the energies of radiation are not frequencies themselves but rather they are differences in frequencies with respect to the corresponding standard substance which has highly shielded protons. This frequency difference is called the **chemical shift**. In most cases the standard used for measuring chemical shifts is **tetramethylsilane**, $(CH_3)_4Si$ (**TMS**). Frequency differences relative to TMS are small, only millionth parts of the frequencies themselves. Therefore the chemical shifts are measured in units called **ppm** (parts per million). Chemical shifts for the most important groups containing hydrogen are listed in the table below.

Proton	Chemical shift δ in ppm	Proton	Chemical shift δ in ppm
H−C−R	0,9 - 1,8	H⧸ =⧹	4,5 - 6,5
H−C−C=C	1,6 - 2,8	H−≡−	2,5
H−C−C—	2,1 - 2,5	H−⬡	6,5 - 8,5
H−C−NR	2,2 - 2,9	H−C=O	9 - 10
H−C−Cl	3,1 - 4,2	H−OR	0,5 - 5
H−C−O	3,2 - 3,7	HOC=O—	10 - 13

The ^1H nmr spectrum of ethanol constructed from the data for chemical shifts in the table above is represented in the next figure. Since an ethanol molecule consists of three different groups containing hydrogen (CH_3, CH_2, and OH) three signals appear in the spectrum.

Chemical shift in **ppm** units

The intensities of these signals depend on the number of hydrogen atoms in the corresponding group, so the ratio of intensities of signals is 3:2:1 for CH_3, CH_2 and OH. It can be seen from the figure, that the spectrum of ethanol is even more complex. The magnetic moments of hydrogen nuclei are somewhat coupled between each other, because of the influence of the electron cloud. Such coupling between protons is detected from the splitting of spectral lines (four lines for the CH_2 and three lines for the CH_3 group). The splitting of the signal is called the **multiplet**; it depends on the number of hydrogen atoms in the neighboring group. If neighboring group has n hydrogens the signal is split into $n + 1$ lines. These multiplets are named according to the number of lines: two lines form a **doublet**, three lines a **triplet** and four lines a **quartet**. In the above spectrum of ethanol the CH_3 group signal is split into a triplet because the neighboring CH_2 group has two hydrogens ($n = 2$), and the CH_2 group is split into a quartet because its neighboring CH_3 group has three hydrogens ($n = 3$). Hydrogen in the OH group exhibits special behavior which will be discussed later in this book and its coupling with the neighboring groups is not simple.

Chapter 4
Alkenes and Alkynes

The simplest alkene is **ethene**, the compound which has previously been called "ethylene". Ethene is produced in large quantities because it is one of the most important substances in the production of a wide variety of organic compounds and technological materials. The parent alkyne is **ethyne** which is marketed under the trade name of "acetylene". This substance is also a very important industrial product. Both ethene and ethyne are flammable; ethyne mixed with oxygen produces a very hot flame so acetylene is used for gas welding. Ethyne can be easily prepared from calcium carbide and water:

Ethene
(Ethylene)

Ethyne
(Acetylene)

PREPARATION OF ETHYNE

$$CaC_2 \ + \ H_2O \ \longrightarrow \ C_2H_2 \ + \ CaO$$

Calcium carbide   Ethyne

4.1 Constitution and Nomenclature

In contrast to the relatively unreactive alkanes, alkenes and alkynes are chemically reactive because they have double and triple carbon-carbon bonds as functional groups. In the IUPAC nomenclature the names of alkenes end with the suffix **-ene** and the position of the double bond is labeled by number of the C-atom on which this double bond begins. In analogy with alkanes with branched chains, the root of the word is the name of the longest chain of C-atoms and the atom numbering is arranged

H. Vančik, *Basic Organic Chemistry for the Life Sciences*,
DOI 10.1007/978-3-319-07605-8_4, © Springer International Publishing Switzerland 2014

so that C-atom with the double bond has the smallest possible number. The rules for naming alkynes are the same as for alkenes with the exception that the suffix is **-yne**.

$$CH_3CH=CHCH_2CH_3$$
1   2   3   4   5

$$CH_3CH=CHCH_2CH_3$$
5   4   3   2   1

Pent-2-ene   **Pent-3-ene**

CORRECT   **INCORRECT**

$$HC\equiv CCH_2CH_3$$ $$CH_3CH_2C\equiv CCH_2CH_3$$

But-1-yne   **Hex-3-yne**

simpler:

Butyne

If the alkene molecule has two or more double bonds the names are formed in accordance with their number, for instance **dienes, trienes, polyenes** etc. If single and double bonds alternate, such alkenes are named **conjugated alkenes**. These structures are responsible for the special properties of conjugated compounds. For example, alkenes which have four or more conjugated double bonds can be colored because they absorb light in the visible region. Double bonds in the molecule can also be located on the neighboring atoms. Such bonds are called **cumulated** and the corresponding compounds are **cumulenes**. The simplest cumulene is **allene**:

CONJUGATED
DOUBLE BONDS

$$H_2C=C=CH_2$$ CUMULATED
Allene   DOUBLE BONDS

Cumulenes with ten or more cumulated carbon atoms have been detected in the interstellar space and some theories propose that such molecules played very important role in the formation of organic molecules during chemical evolution.

4.2   Configuration of Alkenes

Since carbon atoms in alkenes are interconnected with double bonds, every C-atom has three neighbors. As we know from previous chapters, the molecule is more stable if groups surrounding certain central atom are positioned as far as possible from each other (VSEPR method). If the central atom is carbon as in alkenes, its neighbors must lie in the same plane similarly as the C-atoms in graphite. Such planar structure appears not only in alkenes but in all organic molecules that have C-atom with a double bond. In alkynes, the carbon atom has only two neighbors and the alkyne molecule is linear. A similar linear structure is also characteristic of the structure of cumulene molecules.

The largest electron density in planar molecules like alkenes is not located between the two carbon atoms but rather above and below the molecular plane. Such distribution of electron density is rationalized by the molecular orbital model in which the double bond includes π-orbitals as shown in the following figure.

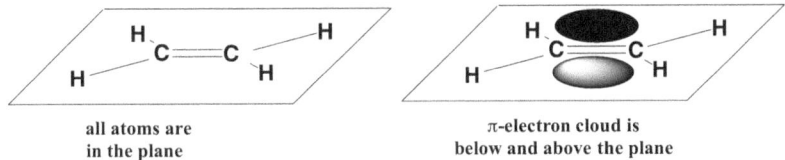

| all atoms are in the plane | π-electron cloud is below and above the plane |

These sterically unhindered π-electron densities are a perfect target for the attack by other particles, atoms or molecules. In the previous discussion about the electronic structure of organic molecules it has been mentioned that the π-orbital is antisymmetric upon the rotation around the bond axis. The consequence of this symmetry property (as well as of the electron density distribution) is that torsion of one CR_2 group relative to other CR_2 group causes breaking of the π-bond. For example, the structure A* which appears after rotation around the C=C bond in but-2-ene (figure below) has a high energy and is very unstable.

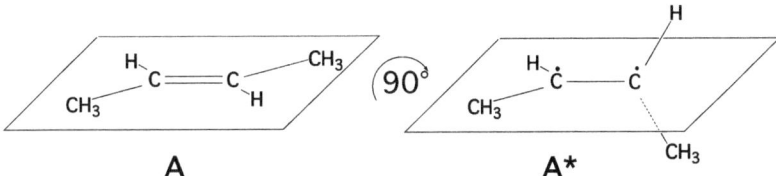

Subsequently, A* may experience rotation around the C=C bond and return back to form A or generate a new molecule B:

Transformation $A \Rightarrow A^* \Rightarrow B$ includes the breaking and re-forming of a chemical bond (in this case a π-bond). The process in which a chemical bond is broken or formed is known as **chemical reaction**. If A is transformed into B by a chemical reaction then A and B are isomers. However, these isomers differ only in the spatial arrangement of functional groups; these isomers have different **configurations**. Isomers which differ from each other only in spatial configuration are called **stereoisomers**.

Stereoisomers must be unequivocally named using appropriate chemical nomenclature. According to the traditional nomenclature such stereoisomers are designated depending on the position of large groups bound to the carbons of the double bond. If the large groups are present on the same side of the double bond, the isomer is labeled by the prefix *cis* and if the groups are situated on opposite sides the prefix is *trans*. We can therefore have *cis*-isomers and *trans*-isomers, respectively.

cis-**But-2-ene**   *trans*-**But-2-ene**

This rule for naming stereoisomers is practical, but still equivocal because most alkene molecules have structures in which it is impossible to decide whether they belong to the *cis*, or to the *trans* configuration. For instance, let us examine the following structure:

What is the criterion for deciding which groups are larger and which smaller, so as to determine which is *cis* and which is *trans* configuration? To solve this problem **Robert Sidney Cahn, Sir Christopher Ingold** and **Vladimir Prelog** have developed the new system of nomenclature for stereoisomers.

Let us apply the **Cahn-Ingold-Prelog** (CUP) system to the molecule shown in the last example via a sequence of steps. In the first step we divide the C-atoms of the double bond by a red broken line as shown in the next figure.

Group with higher priority	Group with higher priority	Group with higher priority

Z-But-2-ene   **E-But-2-ene**

The left and right C-atoms of the double bond are connected to different groups (CH_3 and H on the left C-atom, Br and CH_2Cl on the right C-atom). These groups must be classified by using the **priority rule**. If the groups which have the highest priority are on the same side of the double bond the isomer is labeled with letter Z (by the German word *zusammen*), but if the groups with the highest priority are on the opposite sides, the isomer is labeled with letter E (by German word *entgegen*).

The priorities depend on the atomic number of the atom bound to the C-atom of the double bond. In our example, the left C-atom is bound to atoms C and H. Since the atomic number of C is greater than the atomic number of H, the CH_3 group has priority over H. The right C-atom is connected to atoms Br and C with Br having higher atomic number. Hence, Br has priority over the CH_2Cl group. While in the left isomer, the groups with highest priorities (CH_3 and Br) are on the same side and the structure is the Z-isomer, the groups with the highest priority in the right isomers are on the opposite sides making this the E-isomer. The next example is more complicated because both atoms bound to the C-atom of the double bond are the same (carbons in CH_2OH and CH_2Cl).

In this case, the priority of -CH_2OH and -CH_2Cl groups must be distinguished by inspecting the atomic numbers of atoms bound to carbons in these two functional groups. Carbon atom in -CH_2OH is bound to H, H and O while carbon in -CH_2Cl is bound to H, H and Cl. Since Cl has larger atomic number than O the -CH_2Cl group has priority over the -CH_2OH group and the configurations are:

4.3   Electronic Structure and Reactions of Alkenes

The simplest reaction of alkenes, the $Z - E$ isomerization, was already mentioned in the previous section. The transformation in which one isomer (the **reactant**) is rearranged to give another isomer (the **product**) by rotation of $180°$ requires heating to temperatures above $200\ °C$ in most cases.

However, many *cis-trans* (*Z-E*) isomerizations of natural products are known to occur under mild conditions in living cells. These reactions proceed via different mechanism, utilizing light rather than heat. The well known example is the reaction which occurs in the eye retina. This reaction represents a fundamental step in the process of visual perception.

The molecule of the derivative **retinal** (see figure below) is an alkene with conjugated double bonds. As we have already mentioned the conjugated molecules have the property to absorb visible light. Upon the absorption of a photon the retinal molecule is transformed from the *Z*- to the *E*-isomer. This change in the molecular stereochemistry triggers the signal by which the information about the absorption of a photon is transferred to the nervous system.

Such photoisomerizations are typical of alkenes and can be explained by a shift of one electron from the bonding π to the antibonding π^* orbital. This process is shown in following diagram.

Molecule in the excited state

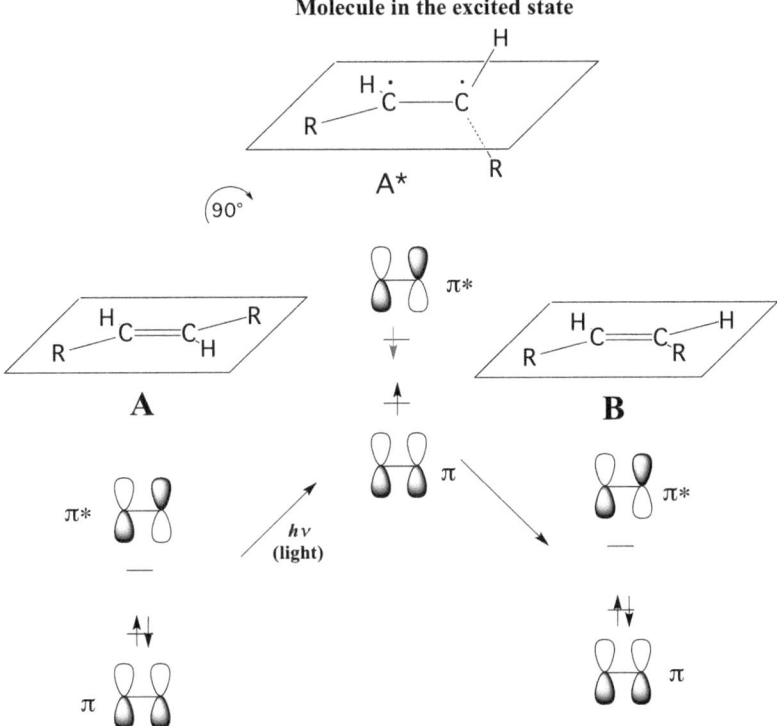

A molecule in which electrons, after absorption of a photon, become shifted to higher orbitals (for instance into π* or σ* orbital) is said to be in an electronic **excited state**. In the example represented in the figure above the excited molecule is described with the structure A*. Since the π* orbital has an antibonding character as we have already discussed, the double bond in A is broken and in A* only the single σ bond remains. Because of the almost free rotation around the single bond (see chapter about conformations) the molecule can be converted to the new *cis* (or *Z*) configuration in which all atoms lie in the same plane and the π-bond can be reformed.

Conjugated alkenes have the property that the energy gap between π and π* orbitals diminishes when the number of conjugated double bonds increases. It is the reason why the alkenes with a small number of conjugated double bonds absorb ultraviolet radiation, while the alkene molecules with a large number of double bonds as for instance retinal, absorb visible light. Molecules absorbing visible light are colored. Hence, retinal is a good sensor for visible light.

4.4   Addition Reactions of Alkenes, and the Concept of Reaction Mechanism

Alkenes are also called **unsaturated hydrocarbons** because their molecules can be transformed into the **saturated** hydrocarbons, the alkanes, by binding molecules of hydrogen. From the formal structural point of view the binding of hydrogen H_2 to the alkene molecule requires several steps. In the first step the H-H bond is broken, followed by the breaking of one π-bond and the formation of two new C-H bonds. Amongst the bonds mentioned, the H-H bond in the hydrogen molecules has the highest energy (434 kJ/mol). Breaking of this bond is possible only under drastic conditions, for instance at high temperature and pressure, which is not always practical in the laboratory

However, it has been found that a hydrogen molecule can be adsorbed on the surface of metals such as palladium or platinum. By such adsorption the electron cloud from the H-H bond is partially redistributed to the metal atoms:

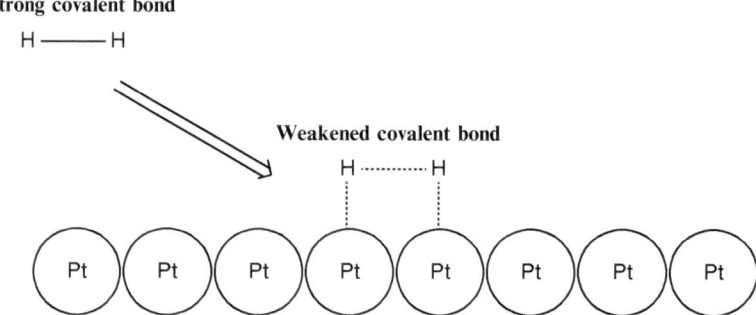

Consequently, the covalent bond between the H-atoms becomes weaker and the hydrogen molecule becomes activated for **addition** to the double bond. On the other hand, π-electrons in the alkene molecule have the highest electron density above and bellow the plane in which carbon atoms lie. Thus the electrons in the alkene molecule are available for the formation of new C-H bonds with activated hydrogen as shown in the next figure. This class of reactions is called **addition** and is typical for unsaturated hydrocarbons. Addition of hydrogen to the unsaturated compounds is called **hydrogenation**. The role of the metal is only to activate the hydrogen molecule so the metal is the **catalyst**. The reaction can in this case be called **catalytic hydrogenation**.

The reaction can be represented by the simple equation:

$$CH_3CH=CHCH_3 \xrightarrow{\text{H}_2\text{, Pt}} CH_3CH_2CH_2CH_3$$

The same reaction is also characteristic for alkynes. The only difference is that alkynes have triple bond and the saturation requires two molecules of hydrogen:

$$CH_3C \equiv CCH_3 \xrightarrow{2\text{H}_2\text{, Pt}} CH_3CH_2CH_2CH_3$$

Let us imagine that in the first step we add one H_2 molecule to alkyne and in the second step another H_2 molecule is added to the remaining double bond of alkene as shown in the next scheme. In organic chemistry the addition of hydrogen is called **reduction** while the removing of hydrogen is called **oxidation**. Oxidation also occurs upon addition of oxygen or halogen.

REDUCTION

$$CH_3C \equiv CCH_3 \underset{-H_2}{\overset{+H_2}{\rightleftarrows}} CH_3CH=CHCH_3 \underset{-H_2}{\overset{+H_2}{\rightleftarrows}} CH_3CH_2CH_2CH_3$$

OXIDATION

4.5 Additions of Hydrogen Halides

Hydrogen halides form alkyl halides by addition to the double bonds of alkenes. For example, hydrogen bromide HBr, can be readily added to propene in the reaction in which hydrogen binds to the carbon atom of the CH_2 group. The resulting product is 2-bromopropane. If HBr is added to 2-methylbut-2-ene the hydrogen binds to the

C-atom of the CH group, yielding 2-bromo-2-methylbutane (see the next scheme). Alternative reactions in which hydrogen binds to the CH group producing 1-bromopropane (propyl bromide) in the first example or to the C=atom yielding 2-bromo-3-methylbutane in the second example are highly unfavored. By carrying out systematic experiments Russian chemist **Vladimir Vasiljevič Markovnikov** found that hydrogen always binds to the alkene carbon which possesses more hydrogens. This empirical finding is known as the **Markovnikov rule**.

$$CH_3-CH=CH_2 \xrightarrow{HBr} CH_3-\underset{\underset{Br}{|}}{CH}-CH_3$$

Propene   2-Bromopropene

$CH_3-CH_2-CH_2Br$

Propyl bromide

NOT
APPEARING

$$CH_3-\underset{\underset{CH_3}{|}}{C}=CHCH_3 \xrightarrow{HBr} CH_3-\underset{\underset{Br}{|}}{\overset{\overset{CH_3}{|}}{C}}-CH_2CH_3$$

2-Methylbut-2-ene   2-Bromo-2-methylbutane

$$CH_3-\overset{\overset{CH_3}{|}}{CH}-\underset{\underset{Br}{|}}{CH}CH_3$$

2-Bromo-3-
methylbutane

Explanation of this rule was not possible in Markovnikov's time because the theory of chemical reactions and electronic structure had not yet been developed. To gain deeper insight into the details of the observed behavior it is necessary to introduce a new chemical concept, the **reaction mechanisms**. The mechanism of chemical reaction describes successive events along the way from **reactants** (the starting substances) to the **products**. What is the nature of these successive steps? One example of the reaction mechanism was already discussed: the reaction of catalytic hydrogenation.

The reaction mechanism theory is connected to the Lewis concept of the electronic structure of molecules. Starting with the Lewis idea that chemical bonds are described as electron pairs, **Sir Robert Robinson** and **Sir Christopher Ingold** have proposed that the reaction mechanism can be explained as the transfer of these electron pairs in such a way that the molecular structure can be rearranged from reactants into products.

Let as apply this concept to the previous examples of additions of hydrogen bromide to 2-methylbut-2-ene. Since the reactant HBr molecule easily dissociates into H^+ and Br^-, the electron deficient H^+ will readily react with the π-electron cloud of the alkene molecule. The H^+ ion is called the **electrophile** because it has affinity for the negatively charged electron clouds. The second ion Br^- is negative and is called the **nucleophile** because it has affinity for the positively charged regions of the molecule of the second reactant. Nucleophiles can be easily recognized because their molecules always possess the nonbonded, electron lone pairs, the **n-electrons**. Several common nucleophiles and electrophiles are listed below:

$$H_2\overset{..}{\underset{..}{O}}:\qquad \overset{-}{:}\overset{..}{\underset{..}{O}}H\qquad \overset{..}{:}\overset{-}{Br}\overset{..}{:}\qquad H_2\overset{..}{S}:\qquad \overset{..}{N}H_3$$

Nucleophiles

$$\overset{+}{H}\qquad\qquad \overset{+}{Na}\qquad\qquad H_3\overset{+}{O}$$

Electrophiles

Now we are ready to explain the mechanism of addition of HBr to the alkene molecule. The electrophile H^+, takes the electron pair from the double bond and binds to the carbon atom. In this process the second carbon atom of the double bond loses one electron and becomes positively charged. Such addition of proton H^+ to the molecule is called the **protonation**.

$$CH_3-CH{=}CH_2 \quad \overset{H}{\overset{+}{}} \longrightarrow \quad CH_3-\overset{+}{CH}-CH_3$$

CARBOCATION

From the upper scheme it is clear that the protonation forms new positively charged molecule which is called **carbocation** and in which the positive charge is mostly localized on the carbon atom. Carbocations are relatively unstable molecules, the **reactive intermediates** which are present for only a short time during some chemical reactions. Since formally either of the alkene carbons could be protonated two carbocations can possibly form. However, experiments and high level quantum chemical calculations have shown that they have different stabilities:

$$CH_3-CH{:}CH_2 \quad \overset{H}{\overset{+}{}} \longrightarrow \quad CH_3-\overset{+}{CH}{-}CH_3 \ \ \text{or} \ \ CH_3{-}CH_2{-}\overset{+}{CH_2}$$

$$\qquad\qquad\qquad\qquad\qquad\qquad \text{stable} \qquad\qquad \text{unstable}$$

CARBOCATION   CARBOCATION

Since only the stable carbocation can survive long enough to react, the Br⁻ nucleophile donates its lone pair to the positively charged carbon atom and binds to it:

$$CH_3-\overset{+}{CH}{-}CH_3 \qquad \overset{..}{:}\overset{-}{Br}\overset{..}{:} \longrightarrow \quad CH_3-\underset{\underset{Br}{|}}{CH}{-}CH_3$$

In summary, the reaction mechanism consists of two steps: (i) the protonation of the double bond resulting in the formation of stable carbocation and (ii) the attack of the nucleophile (Br⁻) on the positive carbon of the same carbocation.

The stability of carbocation depends on the substituents bound to its positively charged carbon. Amongst hydrocarbons, the most stable carbocations are molecules with the positively charged C-atom surrounded by three nonhydrogen neighbors i.e. the **tertiary carbocations**. Less stable are **secondary carbocations** with **primary carbocations** being the least stable. The simplest examples of carbocations are shown in the following scheme.

An important reaction of alkenes is also the addition of water. Since the dissociation of water does not produce a sufficient concentration of protons required for protonation during the first step in the addition mechanism, sulfuric acid must be added for an efficient reaction. After the initial protonation the nucleophilic attack of the water molecule follows, with the subsequent removal of H^+. Addition of water also occurs in accordance with Markovnikov rule because the preferred reactive intermediate is a tertiary carbocation (see the following scheme).

Reagents for additions to alkenes and alkynes can also be halogens. The products of these additions are dihaloalkanes and tetrahaloalkanes, respectively.

The mechanism of halogen addition is more complex. Since there are no protons in the first step it is necessary that the halogen molecule becomes **polarized** when being attached to the double bond. Polarization means that the electron density is redistributed between halogen atoms so that one of the halogen atom becomes partially positive while the other becomes partially negatively charged. In other words the covalent bond becomes polarized. A partially positive halogen atom behaves as an electrophile and binds to one of the alkene C-atoms. The carbocation which then appears is a good target for the addition of the remaining halogen which acts as a nucleophile. In the scheme below the mechanism of addition of chlorine is shown.

Reaction mechanism

$$CH_3CH{=}CHCH_3 \xrightarrow{\overset{\delta+}{Cl}\,\overset{\delta-}{-}Cl} \underset{CI}{CH_3CH{-}CHCH_3}^{+} \xrightarrow{:\!\ddot{Cl}\!:^{-}} \underset{CI\ \ \ \ CI}{CH_3CH{-}CHCH_3}$$

(δ+ and δ- are labels for partial charges)

From the mechanism of catalytic hydrogenation we notice that hydrogen adds to the alkene molecule in such a way that both H-atoms are added to the same side of the molecular plane. The stereochemistry of this reaction is called **syn-addition**. Addition of halogen follows quite different stereochemistry: two halogen atoms approach the alkene molecular plane from the opposite sides. Such stereochemistry is called **anti-addition**. The mechanism is confirmed by the analysis of configurations of products of addition of chlorine to cyclopentene.

anti addition

4.6 Oxidations and Polymerizations of Alkenes

In previous sections we have mentioned that oxidation in organic chemistry includes the removal of hydrogen or the addition of oxygen. Alkenes can easily be oxidized in a reactions with ozone, O_3. The reaction mechanism involves the formation of a reactive intermediate, the ozonide, which subsequently decomposes into two molecules that can be either aldehydes or ketones, depending on the structure of the starting alkene (see the next scheme).

Alkenes can also be oxidized by other reagents such as KMnO$_4$.

Ozonide   ketones

Ozonide   aldehydes

Under high pressure and temperature and in the presence of oxygen or peroxides, ethene is transformed into a solid elastic substance in which the molecules are connected in long carbon chains. This product is called **polyethylene** and the process is **polymerization**. Polymerizations of alkenes are important processes for the industrial production of different technological materials.

Ethene
(Ethylene)   Polyethylene

monomer   polymer

The types and properties of polymeric materials being formed depend on the starting molecule, the **monomer** and on the conditions of polymerization. Almost all industrial polymerizations require the use of an appropriate catalyst. Some polymers and starting monomers are listed in the following scheme.

MONOMERS   **POLYMERS**

4.7 Aromatic Hydrocarbons

Although the aromatic hydrocarbons have a six-membered ring and conjugated double bonds formally characterizing them as unsaturated hydrocarbons, they exhibit distinct chemical behavior, different from the behavior of alkenes. The simplest compound with such properties is benzene C_6H_6, the substance which was discovered by **Michael Faraday** 200 years ago. Derivatives of benzene and other analogous compounds were called aromatic compounds because of their characteristic odor. At that time, determining the molecular structure of benzene represented a problem for chemists because it had been difficult to deduce its constitution by using the structural theory then available. Additions, the most characteristic reactions of alkenes, were never observed with benzene and the catalytic hydrogenation of benzene to cyclohexane, C_6H_{12} is possible only under drastic experimental conditions (high pressure and using specific catalyst). In the middle of nineteenth century, **August Kekulé** (the chemist who is together with **Couper** and **Butlerov** known for the discovery of the property of carbon to form chains) proposed that six carbon atoms in the benzene molecule are connected so as to form a six-membered ring:

Later, chemists isolated other more complex aromatic compounds in which the benzene rings are **condensed** (two rings are condensed if they have a common bond). The most important examples are naphthalene, anthracene and phenanthrene:

| Naphthalene | Anthracene | Phenanthrene |

Aromatic structures can also appear as substituents on the hydrocarbon chains. We then use special prefixes in their nomenclature. For instance, the C_6H_5 group is named **phenyl** and the $C_6H_5CH_2$ group is called **benzyl** as is shown in the next scheme.

Phenyl group   Benzyl group   Triphenylmethane

C_6H_5 $C_6H_5CH_2$

The structure of the benzene molecule proposed by Kekulé was not accepted without criticism. The basic concerns appeared related to the study of substituted derivates of benzene. For example, if the structure includes the alternating double bonds the dimethylbenzene derivative should exist as four isomers:

o-Dimethylbenzene   1,6-Dimethylbenzene   m-Dimethylbenzene
1,2-Dimethylbenzene   1,3-Dimethylbenzene
                                                      p-Dimethylbenzene
                                                      1,4-Dimethylbenzene

The isomers 1,2-dimethylbenzene, 1,3-dimethylbenzene and 1,4-dimethylbenzene also have traditional names: o-dimethylbenzene, m-dimethylbenzene and p-dimethylbenzene (ortho-dimethylbenzene, meta-dimethylbenzene, and para-dimethylbenzene). If the positions of double bonds were **localized**, as proposed by Kekulé, 1,2-dimethylbenzene and 1,6-dimethylbenzene would be different isomers because the bond between carbons C_1 and C_2 is double while the bond between C_1 and C_6 would be a single bond. However, all the experiments have demonstrated that only three isomers exist, hence 1,2-dimethylbenzene and 1,6-dimethylbenzene are identical. Consequently, the classical structure theory which implies that the double bonds are localized is unable to describe the structures of the benzene molecule and other aromatic compounds.

The resolution of this conflict between the structure theory and experimental results was not possible during the Kekulé's time, because the electronic structure of the chemical bond has not been discovered.

It was only during the first decades of twentieth century that **Robinson** and **Arndt** proposed the concept of **delocalized electrons**. Six electrons in three formal double bonds are not localized in these bonds but are equally distributed among all six carbon atoms of the benzene ring. This concept was later supported by the discovery that all carbon-carbon bonds in benzene ring are of equally length (139 pm). Benzene C-C bond length is intermediate between the values for single and double bonds (see previous chapter).

Robinson had also proposed two ways to represent the structures with delocalized electrons. The first representation is the method of resonance structures which we already discussed in this book. By using this method, the benzene molecule is represented by two resonance structures. The other representation proposed by Robinson describes the six delocalized π electrons by plotting a circle inside the six-membered ring.

Resonance
structures

Robinson circle
6 π electrons

Here, it is necessary to point out that the circle corresponds to six π-electrons and cannot be used for the descriptions of naphthalene or other aromatics which possess more than six π-electrons. For the representation of the electronic structures of other aromatic molecules, the method of resonance structures is preferred.

Enumeration of π-electrons present in different aromatic molecules reveals a certain regularity: 6 π-electrons in benzene, 10 π-electrons in naphthalene and 14 π-electrons in anthracene. The numbers 6, 10 and 14 all result from the formula $4N+2$, where N is integer. For 6 π-electrons $N=1$, for 10 π-electrons $N=2$ and for 14 π-electrons $N=3$. Compounds whose molecules possess $4N+2$ delocalized electrons afford greater stability than corresponding conjugated alkenes with planar chains. This additional stability is called the **aromatic effect**.

Within the molecular orbital theory presented earlier in this book, aromatic stabilization is described by π-orbitals, constructed by combining six $2p_z$ atomic orbitals of carbon atoms which are perpendicular to the molecular plane.

Other aromatic molecules have analogous structures described by the combinations of perpendicular $2p_z$ atomic orbitals. In naphthalene the ten $2p_z$ atomic orbitals of carbon are combined into π-orbitals.

The stabilizing aromatic effect explains why the addition reactions characteristic of alkenes do not occur with aromatics compounds. Addition of chlorine to benzene requires special conditions. The reaction can proceed only by irradiation of the mixture of benzene and chlorine with ultraviolet light. The result of this reaction is a mixture of isomers of hexachlorocyclohexane. Some of these isomers known under the commercial name HCH are used as pesticides.

$$\text{benzene} \xrightarrow{h\nu,\ 3Cl_2} \text{Isomer of Hexachlorocyclohexane} + \text{other isomers}$$

**Isomer of
Hexachlorocyclohexane
HCH pesticide**

However, their use is restricted because of negative effects on human health.

Other reactions of aromatic compounds will be discussed later in the section about the electrophilic substitutions.

4.8   Hydrocarbons in Biology

Many unsaturated hydrocarbons are natural products which have an important role in the functioning of living organisms and living communities. The relatively high vapor pressure of simple alkenes is responsible for these compounds being used as communication carriers between organisms. Transfer of information must be based on the emission and detection of molecules with a particular structure. This effect is known as **chemotaxy** and the molecular carriers of information are called **pheromones**. Some compounds that are pollutants can interfere with pheromones in nature and thus disturb communications between organisms.

Ectocarpene,   **Multifidene**

Pheromones of brown algae
Phaeophyceae

Cembrane A,
Pheromone of termite secretion

Chapter 5
Substitutions on Saturated Carbon Atoms

Alkyl Halides, Alcohols, Thiols, Ethers, Amines

In the previous chapters we have discussed two types of reactions, isomerizations and additions. Another major class of organic chemical reactions is a substitution. A substitution in an organic molecule can occur either at the saturated or at the unsaturated carbon atom which is called the **reactive center**. The reaction mechanism of a substitution includes replacement of the **leaving group** with the **incoming group**.

The incoming group can not only be a nucleophile or an electrophile which were already discussed in this book, but also other molecular species such as **radicals**, which are molecules with unpaired electrons. The general layout of a substitution reaction is shown in the scheme below.

H. Vančik, *Basic Organic Chemistry for the Life Sciences*,
DOI 10.1007/978-3-319-07605-8_5, © Springer International Publishing Switzerland 2014

5.1   Radical Substitutions

5.1.1   Alkyl Halides

The substitution reactions in which the incoming and leaving groups are radicals were perhaps the most important processes in early chemical evolution. Because radicals are molecules with one unpaired electron they are exceedingly reactive. In a vacuum, the radical can have a long lifetime because the probability of encountering another particle and undergoing the reaction is very low. This is why cosmic space and outer layers of planetary atmospheres contain large amounts of radicals. However, even under laboratory conditions, some radicals can survive long enough to be detected by spectroscopic methods and be used in practical chemical reactions. Remember that simple inorganic molecules such as NO or NO_2 are also radicals. Some organic radicals are listed in the next scheme.

$$\cdot CH_3 \qquad (CH_3)_3C\cdot \qquad Cl\cdot$$

Methyl radical   *tert*-**Butyl radical**   **Chloro radical**

Phenyl radical   **Benzyl radical**

In most cases organic radicals are products of breaking the carbon-carbon single bond. Since a large amount of energy is required for this process, the reaction in most cases occurs photochemically after the absorption of a quantum of ultraviolet light. To produce a chlorine radical ($Cl\cdot$), a chlorine molecule (Cl_2) must be irradiated with light of the wavelength which corresponds to the Cl-Cl bond energy, which is 243 kJ/mol. Bonds between other halogens are weaker and can be broken by irradiation with light of longer wavelengths. The Br-Br bond energy is 193 kJ/mol and the I-I bond energy is as low as 151 kJ/mol. This is why halogens are convenient reactants in radical substitution reactions which are used for the preparation of **alkyl halides**.

Irradiation of a mixture of methane and chlorine results in the formation of different chlorides of methane. It has been observed that for this photoreaction to proceed it is not necessary to irradiate the reaction mixture continuously during the reaction. It is sufficient to irradiate it for only a short time and the reaction then will continue in the dark. Since the irradiation is used only for triggering the reaction this first step is called **initiation**. As was explained in the previous section this step serves only to break the Cl-Cl bond, i.e. to generate two chlorine atoms (radicals). As shown by the equations presented in the next scheme, the chlorine radical removes one H-atom from the methane molecule which results in the formation of HCl and the methyl radical CH_3·. The methyl radical reacts in the dark with another chlorine molecule producing methyl chloride and another chlorine radical Cl·. The last two reactions which occur in the dark, continue spontaneously and this reaction step is called **propagation**. The system of reactions ends with the recombination of the remaining radicals into stable product molecules. This last step is called **termination** and the overall reaction system is called a **chain reaction**.

$$Cl_2 \xrightarrow{h\nu} 2Cl\bullet \qquad \text{INITIATION}$$

$$Cl\bullet + CH_4 \longrightarrow \bullet CH_3 + HCl$$

$$\text{PROPAGATION}$$

$$\bullet CH_3 + Cl_2 \longrightarrow CH_3Cl + Cl\bullet$$

$$2CH_3\bullet \longrightarrow CH_3CH_3 \qquad \text{TERMINATION}$$

In this reaction a chlorine radical can also react with methyl chloride:

$$Cl\bullet + CH_3Cl \longrightarrow \bullet CH_2Cl + HCl$$

$$\bullet CH_2Cl + Cl_2 \longrightarrow CH_2Cl_2 + Cl\bullet$$

The final product mixture also contains dichloromethane CH_2Cl_2, trichloromethane $CHCl_3$ and tetrachloromethane CCl_4.

Except for the methyl chloride which is gaseous under standard conditions, the chlorides of methane are liquid substances which are used widely in the laboratory, in industry and also in medicine. Trichloromethane, $CHCl_3$, is also known under the traditional name of **chloroform** and can be used as an anesthetic. Tetrachloromethane or carbon tetrachloride is a very good solvent with practical applications in the laboratory, but it must be handled with care because of its carcinogenic properties.

Compounds with different halogens bound to same carbon atom are known as **freons** and they are mostly used as gases in refrigeration systems. Molecules of freons always contain fluorine and other halogens which are mostly chlorine and bromine. Wide application of freons is the consequence of their relatively high vapor pressure and low boiling point. In industry, freons have special nomenclature that is based on the number of carbon, fluorine, and hydrogen atoms. The first number designates the number of C-atoms minus 1, the second number is the number of H-atoms plus 1 and third number gives the number of Cl-atoms. The remaining atoms in the molecule are F-atoms. Some simple freons and their names are given in the following scheme.

Under standard atmospheric conditions, freons behave as chemically inert substances, but they undergo photochemical reactions upon irradiation by ultraviolet light in the higher layers of the stratosphere (at an altitude of 25 km). These photochemical transformations also include reactions with ozone O_3, converting it into the common allotrope of oxygen O_2. Since the stratospheric ozone layer protects Earth's surface from dangerous ultraviolet radiation, its depletion by reactions with freons can have negative ecological effects.

The freon molecule possesses two kinds of chemical bonds, C-F and C-Cl which differ markedly in bond energies. While the C-F bond is strong with a bond energy of 485 kJ/mol, the C-Cl bond is much weaker with a bond energy of 331 kJ/mol. Consequently, irradiation of a freon molecule like for instance CF_3Cl, results in the breaking of the C-Cl bond and yields the following radicals:

$$CF_3Cl \xrightarrow{h\nu} Cl\cdot \; + \; \cdot CF_3$$

The very reactive chlorine radical reacts readily with ozone by removing one oxygen atom and forming the $ClO\cdot$ radical and a molecule of oxygen:

$$Cl\cdot \; + \; O_3 \longrightarrow \; \cdot ClO \; + \; O_2$$

Besides ozone and molecular oxygen, the stratosphere also contains oxygen in the atomic form O. Atomic oxygen reacts with the $ClO\cdot$ radical regenerating the chlorine radical and one O_2 molecule. The chlorine radical can then react again with another ozone molecule. The chlorine atom regenerated in this process allows a single freon molecule to destroy millions of molecules of ozone. The overall process is a chain reaction summarized in the following scheme:

$$CF_3Cl \xrightarrow{h\nu} Cl\cdot \; + \; \cdot CF_3 \qquad \text{INITIATION}$$

- -

$$Cl\cdot \; + \; O_3 \longrightarrow \; \cdot ClO \; + \; O_2 \qquad \text{PROPAGATION}$$
$$\cdot ClO \; + \; O \longrightarrow Cl\cdot \; + \; O_2$$

- -

This knowledge has curtailed the production and use of freons with small molecular masses. However, freons with higher molecular masses do not undergo such chemical reactions and can be used safely in technological applications, for instance in air conditioners.

The undesirable ecological effects have also restricted the applications of **polychlorinated biphenyls (PCBs)**, the substances which are used in the cooling of high voltage electrical transformers. The main disadvantage of the use of PCBs is their slow rate of degradation after use. **Dichlorodiphenyltrichloroethane (DDT)** is a very efficient insecticide, but its use is restricted due to its carcinogenic and neurotoxic effects.

Polychlorinated biphenyl (PCB)

Dichlorodiphenyltrichlorethane (DDT)

5.1.2 Bond Polarity and the Dipole Moment

In the covalent σ-bond, the electron density is highest between the two atomic nuclei. However, if the atoms which form the chemical bond are different, the electron density maximum is shifted closer to the atom which has a higher electronegativity. The property of an atom to attract the electron density is called electronegativity. There are different scales of electronegativity, but the most frequently used are values calculated by Linus Pauling. In principle, the electronegativity of an element increases on going from the left to the right within the period and from bottom to top of the group in the periodic table of elements. For instance, in alkyl halides the halogen atom is more electronegative than carbon so the C-atom becomes partially positively charged and the halogen atom partially negatively charged. Partial positive or negative charges are designated in formulas by δ+ and δ−, respectively. The chemical bond in which the charge is unequally distributed is called the polar bond. In structural formula, the bond polarity is indicated by a vector (bond dipole) as shown in the following scheme.

If the molecule has several polar bonds, the average polarity corresponds to the vector sum of bond dipoles called the **dipole moment** and labeled by the symbol μ.

DIPOLE
MOMENT
μ

The dipole moment is defined as the product of charges and distance between centers of the same charges. The unit for measuring the dipole moment is **debye** (**D**), in the honor of **Peter Debye**, the scientist who made the greatest contribution to research on the polarity of molecules. The molecules with nonzero dipole moments are called **polar molecules**.

$$\mu = Q \times r$$

(**Q** charge in Coulombs, **r** distance between charges in pm)

For the example, if the proton and electron are separated by 100 pm the dipole moment is 4.80 D. This is the limiting value which can be used as a standard for the chemical bond with 100 % ionic character [$(1.60 \times 10^{-29}$ C·m) $(1D/3.336 \times 10^{-30}$ C·m) = 4.80 D]. The C-Cl bond-length is 178 pm and the dipole moment is 1.87 D. If we assume that the C-Cl bond is 100 % ionic, the expected dipole moment would be $\mu = (178/100)$ (4.80 D) = 8.54 D. Since the experimental dipole moment is much smaller, 1.87 D, the character of the C-Cl bond is only 22 % ionic (% of ionic character = $(1.87/8.54) \times 100 = 22$ %). In this way the polarity can be related to the ionic character of the chemical bond.

Since the dipole moment is the vector sum of the polarities of the chemical bonds, it depends on the molecular geometry. Consequently, molecules such as CO_2 or CCl_4, with symmetrically distributed polar bonds, are nonpolar because their total dipole moment equals zero.

As already mentioned, chlorides of methane are very useful solvents. From the knowledge about the polarity of molecules, i.e. their dipole moments, we can distinguish between **polar** and **nonpolar** solvents. The polar solvents are for example CH_2Cl_2 and $CHCl_3$, while tetrachloromethane, CCl_4, is a nonpolar solvent. Polarity of solvents is one of the most important properties, not only for practical laboratory applications, but also for the theory of reaction mechanisms, reactivity and selectivity of organic compounds.

5.2   Reactions of Nucleophilic Substitutions and Eliminations

Nucleophiles are electron rich molecules or ions which may contain for example the electron lone pair, also called the nonbonding electrons. From the building principle of the electronic structure of organic molecules we know that non-bonding n-electrons occupy the highest energy level called the HOMO orbital. Accordingly, these electrons are mostly responsible for the reactivity of organic molecules.

Collision of a nucleophile with an alkyl halide molecule will result in a chemical reaction if the nucleophile attaches to the partially positively charged carbon atom. Since the carbon-bromine bond is polarized, the nucleophile will approach the C-atom bound to the bromine atom (see following scheme). In the given example the incoming nucleophile is the OH⁻ anion and the leaving group is the bromide anion. The product of the reaction of 1-bromopropane with a hydroxide anion is propane-1-ol. One functional group (Br) is replaced by another functional group (OH) and the product belongs to the new class of organic compounds – **alcohols**. For simplicity, in the second equation only the nonbonded electrons which are involved in the reaction mechanism are depicted.

CH₃CH₂CH₂—Br   :OH → **CH₃CH₂CH₂**—OH + :Br⁻

1-Bromopropane   **1-Propanol**

Detailed analysis of this reaction mechanism requires a discussion of the electronic structure of both reactants in the framework of the molecular orbital model. We will avoid this approach because it is outside the scope of this book.

The pertinent question is why Br⁻ as a nucleophile (because it possesses non-bonded electrons) does not react with the product (propane-1-ol) reverting back to 1-bromopropane? Such a reversible reaction is not probable because OH⁻ and Br⁻ have different **nucleophilicities**. While OH⁻ is a stronger nucleophile, Br⁻ is a better leaving group. On the basis of experimental experience and high level quantum-mechanical calculations, chemists have quantified nucleophilicity and arranged it into the **nucleophilic order**. In principle the stronger nucleophile replaces the poorer nucleophile. Nucleophilicity as well as the nucleophilic order are not universal properties because they strongly depend on the reaction environment, especially on the solvent, but in standard chemical practice the following order applies:

$$^-SH > NH_3 > ^-OH > I^- > Br^- > Cl^-$$

5.2.1   Reaction Mechanisms of Nucleophilic Substitutions

The systematic measurements of reaction rates of nucleophilic substitutions have shown that, depending on the structure of reactants, the reaction kinetics can follow either the first or the second order rate law. The chemical reaction rate can be expressed by kinetics equations in which the main parameter is the reaction coefficient (rate constant) k:

REACTION RATE = k [A]

FOR REACTION   A ⟶ products

REACTION RATE = k [A] [B]

FOR REACTION   A + B ⟶ products

While in the first reaction of the scheme above the rate depends only on the concentration of single molecular species A, the rate in the second reaction depends on the concentration of two different molecular species A and B. Therefore, the first reaction follows the **unimolecular mechanism** and the second reaction is described by the **bimolecular mechanism**.

In most nucleophilic substitutions, the substrate with the leaving group bound to a **tertiary** carbon atom follows the **unimolecular** mechanism and the reactant with the leaving group on a **primary** carbon proceeds via the **bimolecular** reaction mechanism.

Christopher Ingold and **Edward D. Hughes**, in the middle of the last century, used such kinetic properties to devise nomenclature for the mechanisms of organic reactions. For the nucleophilic substitutions, Ingold and Hughes have proposed labels S_N1, and S_N2, where **S** and **N** refer to substitution and nucleophilic, respectively. Numbers **1** or **2** are used to designate unimolecular or bimolecular mechanisms, respectively.

The difference in kinetic behavior between the primary and tertiary substrates is a consequence of different underlying reaction mechanisms. Before discussing the details of the mechanisms of nucleophilic substitutions, let us mention the basic principles of the theory of the rates of chemical reaction established by **Henry Eyring**. In a simple chemical reaction, the molecule of the reactant is transformed to the product by passing through a high energy structure which is called the **transition state**. Hence, the Eyring's concept is known under the name of **transition state theory**. For the rearrangement into the transition structure, the reactant molecule must absorb a specific amount of energy called the **activation energy**. The lifetime of the transition structure is very short, around 10^{-13} s; this time is comparable to the period of one molecular vibration. In the previous chapters we have already discussed the nature of molecular vibrations and the corresponding spectroscopic techniques.

The reaction rate expressed through the rate constant k strongly depends on temperature and on the activation energy. This dependence is expressed in the equation given by **Svante Arrhenius** and **Henricus Jacobus van't Hoff**:

$$k = Ae^{-Ea/KT}$$

where k stands for the reaction rate constant, E_a for activation energy, K is **Boltzmann constant** and T is temperature. If the activation energy is low then the transition structure is more stable, the reaction is fast and consequently the reaction constant k is large. This is shown in the next scheme.

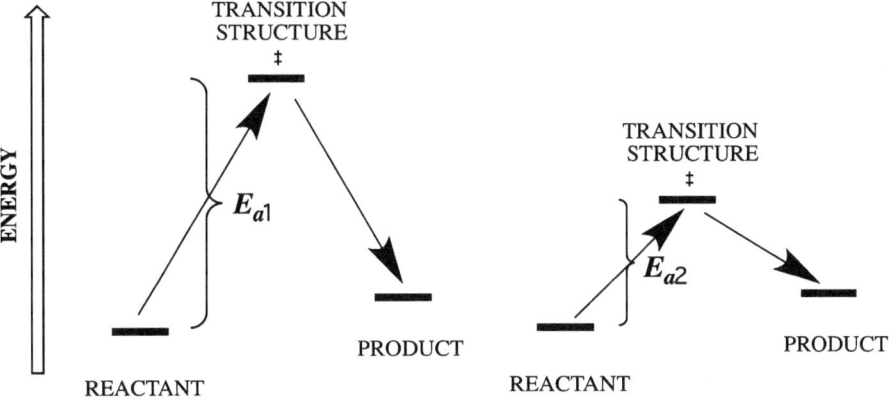

$$E_{a1} > E_{a2} \quad \Rightarrow \quad k_2 > k_1$$

In the two reactions presented, the first one requires a higher activation energy than the second one. Consequently, the second reaction is faster. Such a scheme describes a reactions that follow the S_N2 mechanism. After the collision of the substrate molecule with the nucleophile, provided the molecules have sufficient energy to overcome the activation energy barrier, the transition state is formed. This transition state could be described as a "vibration" in which the incoming nucleophile approaches and the leaving group departs from the reaction center (carbon atom). The breaking of the existing bond and the formation of a new bond are concerted processes, so the S_N2 mechanism is called the concerted mechanism.

The transition state resembles a trigonal bipyramid in which the incoming and leaving groups occupy axial positions and the other three substituents are almost coplanar with the carbon atom. While the central carbon atom holds a partial positive charge the leaving and incoming groups carry a partial negative charge. The rate of this bimolecular reaction depends on the concentrations of the two molecular species, the OH$^-$ ion and 1-bromopropane as shown below:

Nucleophilic substitutions of tertiary substrates follow more a complicated mechanism that consists of two steps. The first step is the dissociation of the bond between the carbon atom and the leaving group. The result of this dissociation is the formation of short-lived reactive intermediate called carbocation. In this book we have already discussed the carbocations in the section about addition reactions. We have mentioned that the most stable are the tertiary carbocations, in which the positive carbon is bound to three neighboring C-atoms. In substrates such as 1-bromopropane in the last example, only the C-Br bond can be cleaved to form a primary carbocation. Since such a primary carbocation is highly unstable, the reaction proceeds via the concerted S_N2 mechanism in which the primary carbocation does not appear.

Let us discuss the S_N1 mechanism in more detail. Dissociation of the C-Cl bond in 2-chloro-2-methylpropane in the first step yields a tertiary carbocation (in this case a *tert*-butyl cation) which as a stable species lives long enough to collide with the OH$^-$ nucleophile and to form the product. It must be pointed out that the stability of the carbocation depends also on the environment. For instance, the cations are more stable in polar than in nonpolar solvents.

Tertiary C-atom

2-Chloro-2-methylpropane
(*tert*-Butylchloride)

tert-Butyl cation

2-Methylpropane-2-ol
(*tert*-Butanol)

As is shown in the scheme above, the reaction consists of two steps: the second step being much faster than the first step. In principle, there are two reactions, one following the other. Such processes are called **consecutive** reactions. The observed rate of such a reaction system corresponds to the rate of the slowest step. This step, in our example the dissociation of the C-Cl bond, is called the **rate-determining step**. Consequently, the observed rate depends only on the concentration of one molecular species, in this case on the concentration of the reactant 2-chloro-2-methylpropane:

$$
Reaction\ rate = k \left[H_3C - \underset{\underset{CH_3}{|}}{\overset{\overset{CH_3}{|}}{C}} - Cl \right]
$$

Unimolecular S_N1 processes can be described in the framework of transition state theory. Since the first, rate-determining step is slower, its activation energy (Ea_1) is higher than the activation energy (Ea_2) for the second, faster step. The complete energy diagram is shown in the next scheme.

TRANSITION STRUCTURE
FOR FIRST STEP

‡

TRANSITION STRUCTURE
FOR SECOND STEP

‡

ENERGY

E_{a1}

E_{a2}

$E_{a1} > E_{a2}$

PRODUCT

CARBOCATION

REACTANT

The reaction mechanisms of S_N1 and S_N2 nucleophilic substitutions are convenient for describing reactions where the reactive carbon atom is either primary or tertiary. Reactions of molecules with a secondary reactive carbon atom usually follow a mechanism that could be described as the combination of these two mechanisms. The presence of an S_N1 or S_N2 mechanism strongly depends on the reaction conditions, mostly on the polarity of the solvent. For instance, polar solvents which stabilize ionic species favor the S_N1 mechanism.

5.2.2   Alcohols

In the previous examples we have demonstrated that nucleophilic substitution in which water and hydroxyl group were used as nucleophiles, yield **alcohols** as products. We note that the carbon atom bound to the OH group can be primary, secondary or tertiary. Accordingly, alcohols can be classified as primary, secondary or tertiary.

PRIMARY	SECONDARY	TERTIARY

Alcohols are named starting with the name of the longest carbon chain and using the suffix **-ol**. Numbering of C-atoms must be such that the carbon on which the OH group is attached has the smallest possible number.

4-Methylheptan-2-ol

Alcohols which have two or more OH groups are called **diols, triols,** *etc.* In the following scheme IUPAC names of some alcohols are given together with their trivial names which are still used, especially in trade or manufacturing.

Ethan-1,2-diol   **Propan-1,2,3-triol**
(Ethylene glycol)   **(Glycerol)**

The simplest primary alcohol is **methanol** or **methyl alcohol,** the substance discovered in eighteenth century by the Irish chemist **Robert Boyle**. Methanol is one of the products of the dry distillation of wood. It is also the product of different biological processes and appears in some alcoholic drinks, which is illegal because of the poisonous nature of methanol. Although the molecule of methanol is small and has a small molecular mass its boiling point, 65 °C, is relatively high. Such a high boiling point is the consequence of hydrogen bonds between methanol molecules. In this property methanol is similar to water and consequently methanol and water can mix in any amount.

99 % pure methanol can be obtained by fraction distillation of the methanol/water mixture. Further distillation does not produce 100 % methanol, because the mixture of methanol and water in a volume ratio of 99/1 has the same composition in the liquid and gas phase. Such mixtures are known as **azeotropes**. Therefore pure or **absolute methanol** cannot be obtained by distillation. The best methods for preparation of absolute alcohols must include chemical reactions with metals such as sodium or magnesium. The products of such reactions are compounds that readily react with traces of water.

Reactions of alcohols with metals yield special salts called **alcoholates** or **alkoxides**. For example, methanol reacts with sodium producing **sodium methanolate** (**sodium methoxide**) with the evolution of hydrogen:

$$2CH_3OH + 2Na \longrightarrow 2CH_3\overset{-}{O}\overset{+}{Na} + H_2$$

Sodium methanolate
or
Sodium methoxide

Sodium methanolate reacts further with water forming sodium hydroxide and methanol. Sodium hydroxide is not soluble in alcohol and precipitates out. The remaining liquid consists only of methanol which can be distilled to get absolute methanol.

$$CH_3\overline{O}\ Na^+ + H_2O \longrightarrow CH_3OH + NaOH$$

For safety reasons, magnesium is used in the preparation of absolute methanol as magnesium is a less reactive metal than sodium.

Ethanol, or **ethyl alcohol**, **CH_3CH_2OH** is one of the basic products of fermentation of fruits, especially wine grapes. Such natural reaction is known as alcoholic fermentation by which the carbohydrate glucose is decomposed into ethanol and carbon dioxide. The reaction is promoted by special biocatalysts called enzymes:

$$C_6H_{12}O_6 \xrightarrow{\text{Enzymes}} 2\,CH_3CH_2OH + 2CO_2$$
$$\text{Glucose} \qquad\qquad\qquad \text{Ethanol}$$

The volume of ethanol in wine produced by fermentation cannot exceed 15 % because the large amount of alcohol deactivates the fermentation enzymes. Distillation of ethanol (boiling point 78 °C) from water yields the azeotrope with 96 % of alcohol. Absolute ethanol can be prepared by adding alkali metals followed by distillation, similarly to the preparation of methanol. **Propanol** forms two isomers, 1-propanol, **$CH_3CH_2CH_2OH$** and propane-2-ol, **$CH_3CH(OH)CH_3$**. These isomers also have the traditional names: ***n*-propanol** and **isopropanol**, respectively. Isopropanol is a very good organic solvent. Both, ethanol and isopropanol are in wide use for disinfection. An excellent liquid for cleaning laboratory glassware is the solution of 5 % of potassium hydroxide in isopropanol.

Ethane-1,2-diol or **ethyleneglycol** belongs to alcohols with more than one hydroxyl group. **Ethyleneglycol** is the main component of antifreeze in car engines. Propane-1,2,3-triol is known under the name **glycerol** and is one of the most important alcohols in living organisms. Glycerol is the main structural component of biologically important compounds lipids; this class of compounds also includes fats. Because of its high viscosity and the relaxing effect on human skin, glycerol has wide use in cosmetics. The reaction of glycerol with nitric acid yields glycerylnitrate, the explosive liquid called **nitroglycerin**. Since pure nitroglycerin is explosive and is so sensitive that cannot be safely transported and handled, its use was very limited until 1867 when Alfred Nobel discovered that the sensitivity of nitroglycerin can be controlled by the addition of some neutral powders. Such powders which adsorb nitroglycerin are used in the explosive called **dynamite**. Taking into account the destructive power of dynamite and its possible military applications, Nobel used the profits from his discovery to establish the prizes awarded to people in the sciences and the arts whose activities promoted world peace and the development of the human society.

$$HOCH_2CHCH_2OH + 3HONO_2 \longrightarrow O_2NOCH_2CHCH_2ONO_2 + 3H_2O$$

OH	ONO$_2$
Glycerol	Glycerol trinitrate Nitroglycerine

Many biologically important compounds are alcohols. Because of their characteristic pleasant odor many alcohols, for the instance **menthol** and **citronellol** are used in the preparation of perfumes. An alcohol with a more complex molecular structure is **cholesterol**, which is a component of cellular membranes and the starting compound for the biosynthesis of various hormones.

Menthol   Citronellol   Cholesterol

The similarity of alcohols to water can be recognized from their behavior in solutions. In the nineteenth century **Svante Arrhenius** discovered that water molecules in the liquid state dissociate into H$^+$ and OH$^-$ ions. We know that H$^+$ ions react with undissociated water molecules and form oxonium ions, H$_3$O$^+$. Analogously, molecules of alcohols can dissociate into H$^+$ (H$_3$O$^+$) and the corresponding alcoholate anion.

$$H_2O \rightleftharpoons \overset{+}{H} + \bar{O}H$$

$$ROH \rightleftharpoons \overset{+}{H} + \bar{O}R$$

Example:

$$CH_3\overset{..}{O}H \rightleftharpoons CH_3\overset{..}{\underset{..}{O}}\bar{} + \overset{+}{H}$$

Methanol   Methanolate
anion

Since the dissociation of water molecules produce equal number of H$^+$ and OH$^-$ ions, the medium is neutral and its pH $= 7$. However, since the dissociation of alcohols yields only H$^+$ ions their solution could be expected to be acidic, i.e. pH < 7. We shall discuss the phenomena of acidity and basicity in more detail.

At the end of eighteenth century, **Antoine Lavoisier** argued that acids are compounds that contain oxygen. In recognition of that Lavoisier named the chemical element *oxygen* using the French word *oxygene*, which means acid.

However, soon after the work of Lavoisier, Scheele prepared an acid which did not contain the element oxygen, namely hydrochloric acid. After the discovery of hydrochloric acid the element essential for the property of acidity became hydrogen. At the beginning of twentieth century, **Johannes Brönsted** and **Thomas M. Lowry** defined acids as substances which are proton donors and bases as proton acceptors.

The equations above showed the behavior of alcohols as acids. In the presence of strong acids, alcohols can also behave as bases:

$$R\ddot{O}H \; + \; H^+ \longrightarrow R\overset{+}{\ddot{O}}H_2$$

Oxonium ion

PROTONATION REACTION

H^+ ions, which originate from a strong acid, bind to the oxygen atom via one of its electron lone pairs. We say that strong acid **protonates** oxygen in the alcohol molecule forming the **oxonium ion**. Accordingly, alcohols can behave either as acids or as bases, depending on the medium.

In general, acidic behavior can be represented by the equation in which the acid is labeled as AH. Anion A^- that remains after removing the proton behaves as a base, because it is in chemical equilibrium and can again accept the proton. Such base is named the **conjugated base**.

$$AH \underset{}{\overset{K_a}{\rightleftharpoons}} A\overset{-}{:} \; + \; H^+$$

Conjugate
base

This equilibrium can be characterized by the equilibrium constant K_a. By convention, the equilibrium constant is defined as the quotient of concentrations of products and reactants:

$$K_a = \frac{[A\overset{-}{:}]\,[H^+]}{[AH]}$$

Because the range of equilibrium constants spans many orders of magnitude it is in practice replaced by the logarithmic parameter pK_a:

$$pK_a = -\log K_a$$

When the acid is more dissociated it is stronger and yields more protons. Hence, if the equilibrium is shifted to right, the constant K_a is larger and the pK_a is smaller. Therefore strong acids are characterized by a small pK_a. The ability of the acid molecule to dissociate depends on the stability of its conjugated base. An alcohol which behaves as a moderate acid is **phenol**, the substance also known as carbolic acid. Because of its acidity, phenol is frequently used as a disinfectant in medicine. Dissociation of the phenol molecule forms a conjugated base called the phenolate anion:

Phenol   Phenolate anion

The stability of the phenolate anion is a consequence of its electron delocalization. Remember that the electron energy is reduced by electron delocalization, as we have demonstrated in the case of the benzene molecule. The existence of electron delocalization in the phenolate ion can be demonstrated using the resonance method. The resonance structures of the phenolate anion are given below:

Resonance structures of phenolate anion

As we can see from the above formulas, the negative charge is not localized on the oxygen atom (as is the case in the methanolate anion CH_3O^-) but rather it is delocalized over the entire benzene ring.

The behavior of alcohol as a base is evident in the chemical reaction in which alkenes are formed. In a strong acid medium, alcohol is protonated at the oxygen atom, forming the oxoniun ion intermediate. The reaction ends with the formation of an alkene by the removal of the water molecule from this oxonium ion. Such reactions, in which some atoms or groups depart from the molecule, are called **elimination reactions**. In most cases the preparation of alkenes from alcohols includes a reaction with concentrated sulfuric acid, which is a sufficiently strong protonating agent.

$$CH_3CH_2CH_2\overset{..}{\overset{..}{O}}H \xrightarrow{\text{H}^+, \text{ H}_2SO_4} CH_3CH_2CH_2\overset{+}{\overset{..}{O}}H_2 \xrightarrow{-\text{ H}_2O} CH_3CH=CH_2$$

Propanol   **Propene**

ELIMINATION REACTION

In the case of tertiary alcohols with different hydrocarbon branches we may expect three different elimination products as shown in the next scheme. However, Russian chemist **Aleksandr Zajtsev** who lived in the nineteenth century used systematic experimental studies to show that, out of the three possible products, only one appears. Zajtsev proposed the rule according to which the preferred product is the one in which the carbons of the double bond have the largest number of alkyl substituents. We say that the reaction obeys **Zajtsev rule**.

2,3-Dimethylpentt-2-ene   APPEARS

2,3-Dimethylpentan-3-ol

3,4-Dimethylpent-2-ene   NOT APPEARS

2-Ethyl-3-methylbutene   NOT APPEARS

5.2.3   Ethers

Alcoholates (alkoxides), which were already mentioned in the discussion on the reactions of alcohols with metals, behave as strong bases and are good nucleophiles. In the middle of nineteenth century **Alexander William Williamson** succeeded in preparing ethers in the reactions of alkoxides with primary alkyl halides. Later it was demonstrated that these reactions follow the S_N2 mechanism.

$$CH_3CH_2\ddot{O}:Na^+ \;+\; CH_3CH_2\!-\!Br \;\longrightarrow\; CH_3CH_2\!-\!O\!-\!CH_2CH_3 \;+\; NaBr$$

Sodium ethanolate
or
Sodium ethoxide

Diethyl ether

However, the reactions with secondary and tertiary alkyl halides yield alkenes, because the elimination reaction is preferred over a substitution. In contrast to the previously mentioned elimination of water from alcohol initiated by a strong acid, the **elimination of hydrogen halide** requires a strong base (in this case sodium ethoxide).

$$CH_3CH_2\ddot{O}:Na^+ \;+\; H_3C\!-\!\underset{\underset{H}{|}}{\overset{\overset{Br}{|}}{C}}\!-\!\underset{\underset{CH_3}{|}}{\overset{\overset{CH_3}{|}}{C}}\!-\!H \;\xrightarrow{-\,HBr}\; \overset{CH_3}{\underset{H}{C}}\!\!=\!\!\overset{CH_3}{\underset{CH_3}{C}} \;+\; CH_3CH_2OH \;+\; NaBr$$

Sodium ethoxide
base

The base, as proton acceptor, **deprotonates** alkyl halide causing the removal of a halide anion. The reaction proceeds via the transition state in which both C-H and C-Br bonds break **simultaneously**. Such reactions are called **concerted**. The reaction is bimolecular because its rate depends on the concentrations of two molecular species, alkyl-bromide and the ethoxy anion. This mechanism is called the **E2 elimination mechanism**.

If the leaving group is bound to a tertiary carbon atom, the slow reaction step involves dissociation of the C-Br bond and the formation of a tertiary carbocation. In that case the reaction rate depends only on the concentration of one molecular species, the alkyl bromide in our example. Accordingly, the reaction is unimolecular and such mechanism is called **E1 elimination**.

E2 MECHANISM

TRANSITION STATE

$$\downarrow \begin{array}{l} -\,HBr \\ -\,CH_3CH_2OH \end{array}$$

$$\overset{CH_3}{\underset{H}{C}}\!\!=\!\!\overset{CH_3}{\underset{CH_3}{C}}$$

E1 MECHANISM

$$
\begin{array}{ccc}
\underset{\substack{| \\ CH_3}}{\overset{\substack{CH_3 \\ |}}{CH_3\,CH_2\,C}}\!\!-\!Br
& \xrightarrow{\;\text{SPORO}\;} &
\underset{\substack{\diagdown \\ CH_3}}{\overset{\substack{CH_3 \\ \diagup}}{CH_3\,CH_2\,C}}\!+
& \xrightarrow[\;-\,H^+\;]{\;\text{BAZA}\;} &
\underset{H}{\overset{CH_3}{{}^{\diagdown}C}}\!=\!\underset{CH_3}{\overset{CH_3}{C^{\diagup}}}
\end{array}
$$

CARBOCATION

Ethers and alcohols can be regarded as structural analogs of a water molecule. While in alcohols only one of the H-atoms is replaced with alkyl group, in ethers both hydrogens are replaced with alkyl groups (see the next scheme). In the nineteenth century Williamson used such analogy to derive the concept of chemical constitution. This first structural theory was known as the **type theory** in which alcohols and ethers belonged to the water structural type. This was the beginning of the structural formulas which are also used today.

$$
\underset{\textbf{Water}}{H\overset{\displaystyle O}{\diagup\;\diagdown}H}
\qquad
\underset{\textbf{Alcohol}}{R\overset{\displaystyle O}{\diagup\;\diagdown}H}
\qquad
\underset{\textbf{Ether}}{R\overset{\displaystyle O}{\diagup\;\diagdown}R'}
$$

STRUCTURAL ANALOGIES

While alcohols have high boiling points because of hydrogen bonds, ethers do not have OH groups and cannot form molecular aggregates. This is why ethers have low boiling points. Dimethyl-ether is gaseous under the standard conditions and diethyl-ether is liquid with the boiling point as low as 32 °C. The similarity and analogy between ether molecules and water is also evident in their molecular polarities. Because of their molecular polarity, ethers are good solvents for chemical reactions that occur through polar reaction intermediates. Ethers are partially miscible with water. For instance, diethyl-ether can mix with 8 % of water. For this reason ether as a solvent for reactions must be dried by using standard methods for the removal of water.

Ether molecules with several ether groups are called oligoethers or polyethers. Some of them have cyclic structures, for instance furan, tetrahydrofuran, pyran or dioxane.

Furan   Tetrhydrofuran   Pyran   Tetrahydropyran   Dioxane

Cyclic ethers with three- and four-membered rings represent a special group of ethers; the epoxides (oxiranes) and oxetanes, respectively. These compounds are very reactive and have important uses in industry. The most basic cyclic ether, oxirane, ethylene-oxide is used for artificial maturation of vegetables.

Epokside
Oxirane   **Oxetane**
Ethylene oxide

Other epoxide compounds are used as starting materials in the production of polymeric materials. The oligomer represented in the following scheme is one of the components of **epoxide resins**.

n = 20 - 25 monomeric units

Component of epoxide resin

Cyclic ethers with several oxygen atoms in the ring are of special chemical importance. The simplest, dioxane, is routinely used as a solvent in organic synthesis. Ring containing ethers with five or six oxygen atoms afford special chemical properties. Inspection of the structure of the molecules represented in the formulas below shows that oxygen atoms form a cavity which is negatively charged because of the dense electron clouds of electron lone pairs. Such molecules resemble crowns and are known as **crown-ethers**.

18-Crown-6

These ethers are named on the basis of the number of O-atoms and the size of the ring. For instance, the molecule in the scheme above is named **18-chown-6** because the ring has 18 atoms and 6 oxygens. The cavity formed by the O-atoms is large enough for metal ions such as K^+ to enter. Crown ethers with cavities of different sizes can capture metal ions of different ionic radii. The synthesis of crown ethers heralded the rise of a new branch of chemistry, **supramolecular chemistry**, which investigates complex molecular aggregates.

5.2.4   Thiols and Sulfides

Sulfur analogs of alcohols and ethers are **thiols** or **mercaptanes** and **sulfides**, respectively. According to the nomenclature rules, thiols are named following the same rules as for alcohols, the only difference being that the suffix **-ol** is replaced with **-thiol**. Most of these substances can be recognized by their unpleasant odor. **Ethanethiol** is used as an additive to natural gas since its odor serves as an indicator of a gas leak in home installations. Pure natural gas is odorless. Some animals such as skunks release an unpleasant odor which originates from thiols shown in the next scheme. However, some thiols have a pleasant odor, for instance the components of the aroma of coffee.

5.2.5   Amines

The molecule of ammonia has an electron lone pair on the nitrogen atom and is a strong nucleophile. In the reaction with alkyl halides it yields **amines**. The reaction mechanism includes removing hydrogen halide as the leaving group.

$$CH_3CH_2CH_2CH_2-Cl \quad \overset{..}{N}H_3 \longrightarrow CH_3CH_2CH_2CH_2-\overset{+}{N}H_3 \; Cl^-$$

$$\downarrow - HCl$$

$$CH_3CH_2CH_2CH_2NH_2$$

Butane amine

As alcohols and ethers are analogs of water, amines can be considered as analogs of ammonia. The basic chemical properties of amines are consistent with this analogy. Any of the three hydrogen atoms in ammonia can be replaced with an alkyl group. Depending on the number of hydrogens substituted, three types of amines can be formed. In the primary amine, only one H-atom is substituted, in the secondary amine, two hydrogens are substituted and in the tertiary amine, three hydrogen atoms are replaced with the alkyl group.

	PRIMARY AMINE	SECONDARY AMINE	TERTIARY AMINE

$$H-\overset{..}{\underset{|}{N}}-H \qquad H_3C-\overset{..}{\underset{|}{N}}-H \qquad H_3C-\overset{..}{\underset{|}{N}}-CH_2CH_3 \qquad H_3C-\overset{..}{\underset{|}{N}}-CH_2CH_3$$
$$H \qquad\qquad H \qquad\qquad H \qquad\qquad CH_2CH_3$$

Ammonia   **Methylamine**   **Ethylmethylamien**   **Diethylmethylamine**

Amines with four substituents are known as **quaternary ammonium salts** and were discovered by **Alexander von Hofmann** during his work on the determination of the structures of natural products. Quaternary ammonium salts can be prepared by reactions of amines with alkyl iodides, mostly with methyl iodide.

$$\left[\text{⬡}-CH_2-\overset{\overset{CH_3}{|}}{\underset{\underset{CH_3}{|}}{\overset{+}{N}}}-CH_3 \right] \; I^-$$

Benzyltrimethylammonium iodide

QUATERNARY AMMONIUM SALT

Since primary amines can be obtained from ammonia by nucleophilic substitution, the secondary amines can be prepared by using a primary amine instead of ammonia, while tertiary amines are accessible by starting from secondary amines.

$$CH_3CH_2CH_2\overset{..}{N}H_2 \quad CH_3CCH_2CH_2-Cl \xrightarrow[-HCl]{} CH_3CH_2CH_2\overset{..}{N}HCH_2CH_2CH_2CH_3$$

Butylpropylamine

$$CH_3CH_2\overset{..}{N}HCH_2CH_3 + CH_3CH_2Cl \xrightarrow[- HCl]{}$$

CH₃CH₂ \N—CH₂CH₃ / CH₃CH₂

Triethylamine

In water solution, ammonia behaves as a base because the molecule can accept a proton at its nitrogen lone pair. All amines behave similarly to ammonia: they are **organic bases** because they act as **proton-acceptors**.

$$\overset{..}{N}H_3 + H_2O \longrightarrow NH_4OH$$

$$\overset{..}{N}H_3 + \overset{+}{H} + \overset{-}{O}H \longrightarrow \overset{+}{N}H_4\overset{-}{O}H$$

$$CH_3CH_2\overset{..}{N}H_2 + H_2O \longrightarrow CH_3CH_2\overset{+}{N}H_3 + \overset{-}{O}H$$
Ethyl amine   **Ethylammonium ion**

For example, the protonation of ammonia yields the ammonium ion and the protonation of ethyl amine forms the ethylammonium ion. The behavior of bases can be described by a chemical equilibrium and the corresponding constant K_b, similarly as for acids. Analogously, the negative logarithm of K_b (pK_b) is used as a measure of basicity. As pK_b gets smaller, the base is stronger.

$$B: + \overset{+}{H} \underset{}{\overset{K_b}{\rightleftharpoons}} B\overset{+}{H} \qquad K_b = \frac{\left[BH^+\right]}{\left[B:\right]\left[H^+\right]}$$
Conjugate acid

$$pK_b = -\log K_b$$

Acids and bases can also be considered from another point of view. In this view **acids are electron acceptors** while bases are **electron donors. Lewis** has used this **definition** to extend the concept of acidity and basicity to the compounds which don't have hydrogen. The molecule of amine is a stronger base if the electron lone pair on nitrogen is more localized. Localization or delocalization can be estimated by writing the appropriate resonance structures. Electrons are localized if only one resonance formula can be drawn. In principle, if the electron pair is delocalized over a large part of the molecule, the electron energy is lower, the system is more stable and does not require the addition of a proton.

Let us elaborate this concept in case of the basicity of **aniline** and **pyridine**. The nonbonded electron pair at the nitrogen atom is delocalized over the entire molecule, as is evident from the resonance structures shown below.

Resonance structures of aniline

In the pyridine molecule the nitrogen electron pair cannot be delocalized because it occupies the $2p_x$ orbital which is perpendicular to the $2p_z$ orbitals of the aromatic ring.

Pyridine

Hence this electron pair remains localized on the nitrogen atom, its electron energy is higher and its protonation is favored. For this reason pyridine is a stronger base than aniline.

Chapter 6
Nucleophilic Additions

Compounds with a Carbonyl Group

Organic compounds with a carbonyl group are not only numerous but also represent excellent substrates for various chemical reactions. This due to the nature of the carbonyl group and its electronic structure. Since the carbonyl group has oxygen as the heteroatom bound by double bond to the carbon atom, its electrons are distributed amongst σ, π and n orbitals.

The C=O group is polar because of the high electron affinity of the oxygen atom. Consequently, the carbon atom carries partial positive and the oxygen atom partial negative charge. This polarity is evident in the resonance structures below. The dipole moment of the carbonyl group is relatively high and varies between 1.7 and 3.6 D.

Polarity of
carbonyl group

Resonance structures of
carbonyl group

Nucleophiles can attack the partially positive carbon atom in the carbonyl group (see following scheme). The formation of the new chemical bond between the nucleophile and the carbonyl carbon induces an electron shift from the carbon-oxygen

H. Vančik, *Basic Organic Chemistry for the Life Sciences*,
DOI 10.1007/978-3-319-07605-8_6, © Springer International Publishing Switzerland 2014

double bond to the oxygen atom. The result is a reaction intermediate with the negative charge on oxygen. This mechanism is universal and describes all the reactions called **nucleophilic additions to the carbonyl group**.

6.1   Aldehydes and Ketones

Aldehydes appear as products of oxidation of primary alcohols while ketones are products of the oxidation of secondary alcohols. Since oxidation is recognized as the reaction in which hydrogen is removed from the molecule, the removal of a hydrogen molecule from the primary alcohol yields a compound that has historically been called *alcohol dehydrogenatus*. The word **aldehyde** is derived from the first few letters of this historic name.

Besides being prepared by oxidation, aldehydes and ketones can also be prepared by reactions in which the first step includes the addition of water to the triple bond of the alkyne molecule. The first intermediate, the unsaturated alcohol (**enol**) is unstable and undergoes isomerization to the stable ketone. This type of reaction in which one isomer is transformed to another is called **rearrangement**. The older name for this molecular rearrangement is **tautomerism** and this special case is called the **keto-enol tautomerism**.

KETO-ENOL TAUTOMERISM

The constitutional difference between aldehydes and ketones lies in the substituents at the carbonyl carbon atom. While in aldehydes one of the substituents is always hydrogen in ketones both substituents are alkyl groups. Therefore, the chemical properties of aldehydes and ketones, especially in nucleophilic addition reactions are similar. The simplest aldehyde is **methanal** or **formaldehyde** and the parent ketone is **propan-2-one** or **acetone**.

| | | **Methanal** (Formaldehyde) | **Propan-2-one** (2-Oxopropane) (Acetone) |
| Aldehide | Ketone | | |

In the systematic nomenclature the names of aldehydes are formed with the suffix **-al**. In complex molecules the aldehyde group is designated by the suffix **-carbaldehyde**. The names of ketones include the suffix **-one** and the numerical label of the carbonyl carbon atom. The numbering of carbons must be such that each one has the smallest possible number. Ketones can also be named by using the prefix **oxo-** and the numerical label of the carbonyl group. The prefix oxo- is used when the compound possesses both aldehyde and ketone functional groups. The longest chain for naming such molecules is defined by the position of the aldehyde, rather than the keto group.

Ethanal
(Acetaldehyde)   4,4-Dimethylpentanal   Cyclopentylcarbaldehide

4-Chloro-2-cyclohexencarbaldehyde   2,4-Dimethyl-3-oxohexanal

In the following scheme we list structures and the traditional names of some of the important and well-known aldehydes and ketones. Most of them are natural products, which shall be discussed in the subsequent chapters of this book.

Formaldehyde · Benzaldehyde · Acrolein · Acetone

Camphor · Testosterone · Progesterone

In water solution, aldehydes react with water molecules as nucleophiles. The OH^- nucleophile attacks the carbonyl carbon forming a new C-O bond and shifting the electron pair from the double C=O bond to the oxygen atom. In the reactive intermediate, the former carbonyl oxygen becomes negatively charged, as is shown in the next scheme. This negative oxygen is easily protonated and converted to a hydroxyl group. The final product is called a diol, in which both OH groups are bound to the same carbon atom. Such diols are called **geminal diols**. The reaction is reversible and the elimination of a water molecule yields the aldehyde reactants.

In general, the reaction can be represented as:

The position of equilibrium depends on the nature of the substituent R. In some aldehydes such as **chloral**, CCl_3CHO or formaldehyde HCHO, the equilibrium is shifted towards the diol. In its diol form chloral is known as **chloral hydrate** and serves as a disinfectant. The trade name of formaldehyde is **formol** which is actually the water solution of formaldehyde.

Under the standard conditions formaldehyde is gaseous. However, its molecules can interact with each other by a mechanism similar to nucleophilic attack. The oxygen of the carbonyl group behaves as the nucleophilic atom (see the next scheme). The product of such a reaction is a polymer, the white powder called **paraformaldehyde**. Hence, formaldehyde can be provided either in the form of the water solution formol or as paraformaldehyde powder. Pure formaldehyde gas for use in chemical reactions can be prepared by heating paraformaldehyde under vacuum. Similarly, the molecules of acetaldehyde can condense into cyclic trimers giving the substance known as **paraldehyde**.

Paraformaldehhyde

Acetaldehyde

Paraldehyde

Alcoholate anions in alcohol solutions react with aldehydes and ketones forming **ketals** and **acetals**, or **hemiacetals** and **hemiketals**. The rates of nucleophilic additions on the carbonyl group are enhanced by protonation of the carbonyl oxygen, so these reactions are **acid catalysed**.

Butan-2-one

2-Methoxybutan-2-ol
HEMIKETAL

2,2-Dimethoxybutane
KETAL

Propanal

1-Methoxy-1-propanol
HEMIACETAL

1,1-Dimethoxypropan
ACETAL

Acetals and ketals are unstable in acidic solutions and undergo hydrolysis reverting to the original aldehydes and ketones.

The mechanism analogous to acid catalysed nucleophilic addition to the carbonyl group applies to the reactions of aldehydes with different nucleophiles like halide or nitrile ions, yielding **halohydrins** or **cyanohydrins**.

The preparation of cyanohydrins is an important reaction in organic synthesis because it is used to lengthen the hydrocarbon chain by one C-atom: the molecule of 1-cyanopropanol in the scheme above has one carbon atom more than the starting propanal. Some cyanohydrins are natural products, especially the derivatives of **benzaldehyde**, which are poisonous and used by particular insects for their defense. The toxic activity of these compounds is based on the hydrolysis of cyanohydrins into benzaldehyde and highly toxic HCN.

6.1.1 Carbon as a Nucleophile

As we mentioned at the beginning of this book, the large diversity of hydrocarbon compounds is the foundation of organic and biological evolution. This has been proposed by considering the large number of possible isomers. Aldehydes and ketones undergo simple transformations by which high structural diversity can be obtained, starting with only a few simple compounds. In alkaline medium, carbonyl compounds can act as acids, i.e. they behave as proton donors (see following scheme). A stable conjugate base can appear only after removing the proton from the C-atom which is directly bound to the carbonyl group.

Stability of the conjugated base arises from the electron delocalization that can be identified by using the resonance structure method. Acids in which the proton is removed from the carbon atom are called **carbon acids** and their conjugate base is called an **enolate ion**.

$$\left[CH_3CH_2CH_2\overset{-}{\ddot{C}}H-\overset{\overset{\displaystyle \ddot{O}:}{\|}}{C}-H \quad \longleftrightarrow \quad CH_3CH_2CH_2CH=\overset{\overset{\displaystyle :\ddot{O}:^-}{}}{C}-H \right]$$

The anions which would be generated by removing a proton from the carbon atoms which are not neighbors of the carbonyl group are unstable because their electron pair is localized. No resonance structures can be written in such cases (see the structure below).

$$CH_3CH_2\overset{-}{\underset{..}{\ddot{C}}}HCH_2-\overset{\overset{\displaystyle \ddot{O}:}{\|}}{C}-H$$

Since under basic conditions the aldehyde molecules partially dissociate, their mixture contains two molecular species, dissociated and undissociated. The dissociated molecule, the enolate ion can behave as nucleophile because of the electron pair on its carbon atom. This nucleophile can attack the carbonyl group of the undissociated aldehyde yielding a product with two functional groups, aldehyde and alcohol. Such compounds are named **aldols** and the corresponding reaction is called **aldol condensation**.

ALDOL

After loosing a molecule of water, aldol is transformed into the corresponding unsaturated aldehyde.

$$CH_3CH_2CH_2CH_2\underset{\underset{CH_3CH_2CH_2}{|}}{\overset{\overset{OH}{|}}{C}H}\overset{\overset{O}{||}}{C}H \xrightarrow{-H_2O} CH_3CH_2CH_2CH_2CH=\underset{\underset{CH_3CH_2CH_2}{|}}{C}\overset{\overset{O}{||}}{C}H$$

ALDOL

The example given is the reaction in which a molecule with five C-atoms in the hydrocarbon chain (pentanal) gets converted to a molecule containing nine atoms in the longest chain. Let us imagine a mixture of two different aldehydes, A and B. Since these aldehydes can form two enolate nucleophiles and two undissociated molecules, the cross reactions must yield four products: A+A, B+B, A+B and B+A. The number of products of aldol condensation follows the n^2 rule where n is the number of different aldehydes used. For instance, the mixture of ten different aldehydes will produce a hundred different products. If unsymmetrical ketones with two different carbons next to the carbonyl group are added to the starting mixture, the diversity of products in the mixture is further enhanced. This example demonstrates how a large number of diverse structures can originate from a small number of starting molecules.

Molecules with carbon as a nucleophilic center also appear in the reactions of carbonyl compounds with organometallic compounds. Such reactions are very useful in organic synthesis. The best known reagents that include such carbon nucleophiles are alkyl- or arylmagnesium halides, are called **Grignard reagents**, in honor of **Victor Grignard**. The reagents can be easily prepared by direct reactions of an alkyl- or aryl halide with the magnesium metal. Since Grignard reagents are like many organometallic compounds, unstable and sensitive to moisture, the reaction must be performed in dry diethyl-ether.

$$CH_3I \xrightarrow[\substack{\text{Dry} \\ \text{Diethyl ether}}]{Mg} CH_3MgI$$

Methylmagnesium iodide

Phenylmagnesium bromide

The carbon-magnesium bond is polar with the partial negative charge on the carbon atom. Consequently, this nucleophilic C-atom attacks the carbon atom of the carbonyl group while the positive magnesium halide residue bonds to oxygen. The intermediate produced in the next reaction step is hydrolyzed to the corresponding alcohol. It is important to observe that the product molecule has one additional methyl group.

Phenylmethyl ketone

2-Phenylpropan-2-ol

Starting with alkyl halides, we can prepare the appropriate Grignard reagents and in combination with different aldehydes or ketones, synthesize a series of compounds with branched hydrocarbon chains.

6.1.2 Condensations with Amines

Aldehydes and ketones react readily with equimolar amounts of primary amines, yielding imines and water. **Imines** (also called **Schiff bases**) can be considered to be amines with carbon-nitrogen double bonds.

Benzanal   *N*-Benzylidene-*N*-ethanamine
                                                    IMINE

6.1.3 Reductions of Aldehydes and Ketones

Since aldehydes and ketones are produced by oxidations of primary and secondary alcohols, it must be possible to achieve the reverse transformation i.e. their reduction back to alcohols. For this reaction we use special reducing reagents, the metal hydrides, because the reduction in organic reactions involves the addition of a hydride ion H⁻. The best known reducing agents are **lithium aluminium hydride**, $LiAlH_4$, **sodium boron hydride,** $NaBH_4$ and **sodium hydride**, NaH. Molecules of $LiAlH_4$ and $NaBH_4$ consist of lithium or sodium cations and AlH_4^- or BH_4^- anions.

The hydride anion (H⁻) behaves as a nucleophile in the reduction of a carbonyl group and binds to the positively charged carbon atom. The remaining AlH_3 is a Lewis acid which, as an electron acceptor, attacks the oxygen. After hydrolysis and release of aluminum hydroxide, the reaction ends with the formation of an alcohol, in this case propanol (see the following scheme).

The reduction of a ketone follows the same mechanism. In the following example, acetone is reduced to propane-2-ol.

6.2   Carboxylic Acids

In organic chemistry, the oxidation process can also imply the addition of oxygen atoms. In such a way the aldehyde group is oxidized to the **carboxyl group** by adding an oxygen atom. The compounds with a carboxyl group belong to the **carboxylic acids**. Oxidative agents for this transformation include sulfuric acid, solutions of potassium permanganate **$KMnO_4$**, potassium bichromate **$K_2Cr_2O_7$** or chromium trioxide **CrO_3**. The following example represents the oxidation of pentanal to the corresponding carboxylic acid.

Justus von Liebig and **Bernhard Tollens** have discovered that milder agents such as silver oxide, **Ag_2O** oxidize aldehydes to carboxylic acids. This reaction, which is characteristic for aldehydes, serves as a specific qualitative test for this class of compounds and is known as the **Tollens test**. Metallic silver being one of the byproducts, precipitates and forms the characteristic silver mirror that can be easily observed.

An alternative method for the preparation of carboxylic acids is the reaction of nitriles with sulfuric acid. This reaction is important in organic synthesis because it can be used to extend the hydrocarbon chain by one carbon atom. Nitriles can be obtained from the corresponding alkyl halides in the reaction with HCN by nucleophilic substitution. Nitrile can easily be oxidized to carboxylic acid.

Many essential natural products that are ubiquitous in living organisms are carboxylic acids. The simplest (parent) carboxylic acid is **formic acid** (*acidum formicum*) which is named **methanoic acid** in the IUPAC nomenclature. This acid is the main component of the poison of ants.

Acetic acid (*acidum aceticum*) with IUPAC name **ethanoic acid** is used as an everyday food additive. **Citric acid**, which is also a natural product, is a polyfunctional compound with three carboxylic and one OH group. A molecule of the simple **oxalic acid** consists of only two carboxylic groups. Oxalic acid salts are frequently present in kidney stones.

Many carboxylic acids such as **cholic acid** or long-chained **fatty acids** are complex molecules. In contrast to the saturated fatty acids, molecules of unsaturated fatty acids always have carbon-carbon double bonds. Interestingly, the molecules of fatty acids, without exception, have even number of C-atoms. This is an indication that the natural pathways for biosynthesis of these classes of compounds are the same. The role of fatty acids and other natural products in living organisms will be discussed later in this book.

Cholic acid

Oleic acid

Linoleic acid

Stearic acid

The rules for naming carboxylic acids include principles of organic and inorganic nomenclature. The name of the acid consists of the name of the longest hydrocarbon chain and the prefix **-oic acid**. In numbering C-atoms of the hydrocarbon chain, the smallest number must be assigned to the carboxyl group. If the carboxyl group is only a substituent, when its carbon atom does not belong to the hydrocarbon chain, the name is formed with the suffix **carboxylic acid** (as cyclohexanecarboxylic acid in the following example).

Hexanoic acid   **4-Oxoexanoic acid**   **Cyclohexanecarboxylic acid**

Carboxylic acids are substances with high boiling points: formic acid 101 °C, acetic acid 118 °C and propanoic acid 141 °C. This is the consequence of dimerization and the formation of molecular aggregates via hydrogen bonds. Even in the gas phase, the molecules can remain in dimeric form.

HYDROGEN BOND

$$\text{H}_3\text{C}-\text{C}\overset{\text{O}----\text{HO}}{\underset{\text{OH}----\text{O}}{\big\langle}}\text{C}-\text{CH}_3$$

**Dimer of the acetic acid molecules
interconnected by hydrogen bonds**

The strong **acidic behavior** of carboxylic acids as good proton donors is the consequence of stabilization of the corresponding conjugated base (**carboxylate anion**). As can be seen from the resonance structures presented below the electrons in the carboxylate anion are delocalized.

$$\text{H}_3\text{C}-\text{C}\overset{\text{O}}{\underset{\text{OH}}{\big\langle}} \rightleftharpoons \text{H}_3\text{C}-\text{C}\overset{\text{O}}{\underset{\text{O}^-}{\big\langle}} + \text{H}^+ \qquad \left[\text{H}_3\text{C}-\text{C}\overset{\ddot{\text{O}}:}{\underset{\ddot{\text{O}}:^-}{\big\langle}} \longleftrightarrow \text{H}_3\text{C}-\text{C}\overset{\ddot{\text{O}}:^-}{\underset{\ddot{\text{O}}:}{\big\langle}} \right]$$

Ethanoic acid   **Ethanoate anion**

RESONANCE STRUCTURES
of ethanoate anion

The structures show that the negative charge is equally distributed on both oxygen atoms. The charge distribution in the molecule can be calculated by using quantum-mechanical methods and molecular modeling. When the carboxylate anion (in our case ethanoate anion) is modeled by such methods we obtain the diagram shown below with the regions of the highest electron density (negative charge) shown in red:

We have already mentioned that the acidity of organic compounds can be expressed by the acidity constant, K_a, or its negative logarithm, pK_a. Remember that stronger acids have lower pK_a values. The relationship between molecular structure and acidity for different carboxylic acids, which is shown in the table below, can be used to study the structure-activity relationships of organic compounds in general.

Carboxylic acid	pK_a
CH_3COOH	4.7
CH_3CH_2COOH	4.9
CH_2FCOOH	2.6
$CH_2ClCOOH$	2.9
$CHCl_2COOH$	1.3
CCl_3COOH	0.9

The first pattern that is evident from these data is the influence of halogens on acidity. As halogens become more electronegative or when their number increases, the carboxylic acid becomes stronger (it has lower pK_a). Since it is not possible to write resonance structures which would include electrons from the halogens, the electron delocalization in such anions must have a different origin. To get a better insight into this alternative mode of charge delocalization, let us discuss a number of additional examples.

pK_a 2,89   4,05   4,53

Acidity amongst the three chlorobutanoic acids decreases as the chlorine atom moves away from the carboxyl group. The weakest is 4-chlorobutanoic acid with a pK_a of 4,53. It is known that since the electronegative halogens attract electron density, their neighboring atoms become partially positively charged. Such effect is most pronounced on the first neighboring C-atom and it gets smaller along the hydrocarbon chain, as shown in the scheme below.

Charge transfer through σ-bond

INDUCTIVE EFFEKT

Negative charge that is concentrated on the Cl-atom induces progressively weaker positive charges on the rest of the chain. Such electron delocalization is achieved through σ-bonds and is called the **inductive effect**. Since the chlorine atom in the series of chlorobutanoic acid anions is located at different distances from the carboxylic group, its inductive stabilizing influence on the anion is also different: the strongest is in 2-chlorobutanoic acid with a pK_a value of 2.89.

In summary, the carboxylic acid is stronger if after the removal of the proton, it can form a more stable conjugated base, the carboxylate anion. The anion is stabilized by electron delocalization over the largest possible region of the molecule. The **charge delocalization** can occur in two ways: through the π-electron system by the **resonance effect** or through the σ-electron network by the **inductive effect**. The resonance effect has already been explained in the section on alcohols when we discussed the acidity of phenol (*vide infra*).

In the analogy with inorganic acids, the carboxylic acids can be neutralized by bases. The names of the salts which are formed after neutralization follow the inorganic nomenclature.

Ethanoic acid
(Acetic acid)   →   **Sodium ethanoate**
                             (Sodium acetate)

Methanoic acid
(Formic acid)   →   **Potassium methanoate**
                             (Potassium formate)

Of special interest are sodium and potassium salts of the long-chained fatty acids, the compounds known under the traditional name of **soaps** (see the next scheme). The chemical behavior of these molecules exhibits a dual nature. The long hydrocarbon chain is typical of the organic component which is insoluble in water. Because hydrocarbon chains repel water molecules, they are called **hydrophobic**. On the other side the carboxylate group and the metal ion are bound by an ionic bond and such compounds are readily soluble in water. They attract water molecules and we call them **hydrophilic**. The molecules of soaps possess a hydrophilic head and a hydrophobic tail.

HYDROPHOBIC TAIL

Potassium stearate
SOAP

HYDROPHYLIC
HEAD

Because of their specific structures, the soap molecules exhibit special behavior in water solutions, they undergo **self-organization**. The molecules form complex structures called **micelles**, in which the hydrophobic tails are oriented towards the middle of the particle and the hydrophilic heads are pointed towards the water solvent.

The medium inside micelles is hydrophobic and behaves as a non-polar organic solvent. Because of their dual nature, the soaps are used for removing fatty impurities, which cannot be washed away with water. The micelle absorbs the impurity into its hydrophobic central part, but the whole particle is hydrophilic and can be easily washed away by water. Similar behavior is shown by detergents, the compounds that differ from soaps by the nature of their hydrophilic head. The most popular detergents are salts of organic sulfates or quaternary ammonium salts (see for example **SDS**, and **CTAB** in the following scheme). The hydrophilic head in SDS is an anion so these **micelles are called anionic**, while the head in CTAB is cationic and these are named **cationic micelles**. Substances whose molecules can form such micellar structures belong to the large class of compounds, the **surfactants**, which are substances of great practical and industrial significance.

Chapter 7
Stereochemistry, Symmetry and Molecular Chirality

Joseph Achille LeBel was one of the most important pupils and collaborators of **Louis Pasteur**. Both of them were interested in the salts of tartaric acid (contained in wine), the tartarates. During the isolation of ammonium-sodium-tartarate, Pasteur and LeBel observed that two kinds of crystals appear (see figure bellow). Although these crystals are equal in shape they differ in such a way as to be mirror picture of one another.

H. Vančik, *Basic Organic Chemistry for the Life Sciences*,
DOI 10.1007/978-3-319-07605-8_7, © Springer International Publishing Switzerland 2014

This is similar to the spatial relationship between the left and the right hand which cannot be superimposed by any possible reorientation in space. Since LeBel's crystals reveal the same property they have been called chiral i.e. similar to the hand (Greek word for hand is χείρ). Almost at the same time, **Jean Baptist Biot**, the physicist who also collaborated with Pasteur, discovered polarized light. Today we know that light is an electromagnetic wave described as a vibration of two perpendicular vectors, electric and magnetic. If both vectors vibrate in fixed planes, the light is **linearly or plane polarized**. Natural visible light is a mixture of numerous linearly polarized waves. Some substances like polymers and certain crystals transmit only one plane of the propagating light. They are called **polarizers**.

POLARIZER

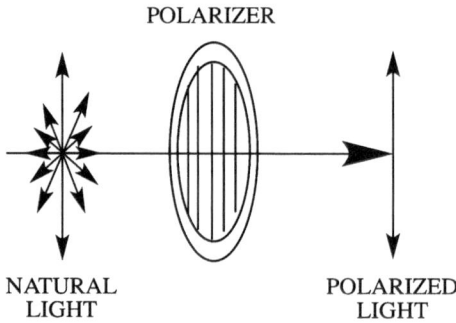

NATURAL   POLARIZED
LIGHT   LIGHT

By passing through the polarizer, the nonpolarized light becomes linearly polarized. If the polarized light passes through two polarizers in sequence, the extent of passage will depend on the relative orientation of the two polarizers. When the polarizing planes in both polarizers are parallel the light will pass through both of them, but if the planes are perpendicular the light cannot pass trough the second polarizer.

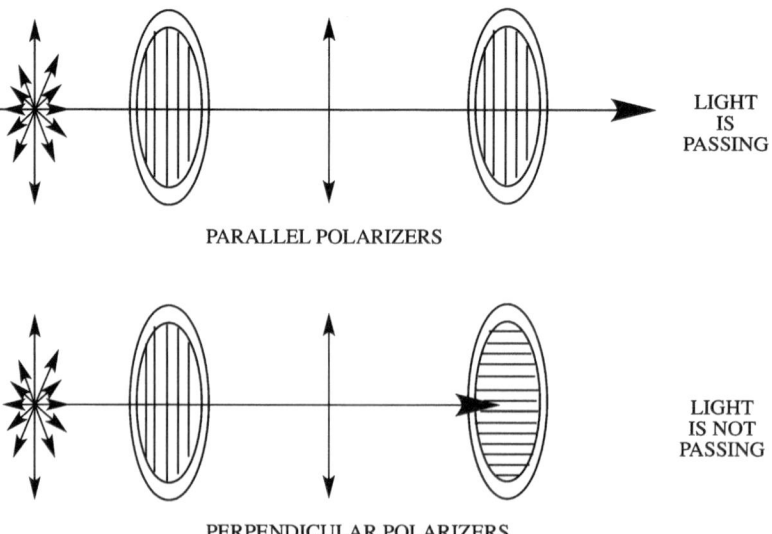

LeBel discovered that if the solution of only one of the two types of crystals (for instance one at the left side of the diagram) is placed between the two polarizers the polarized light passes through both polarizers. However, for this to occur the polarizers are not oriented in parallel, but are rotated with respect to each by a certain angle. The solution of the crystals at the right hand side of the diagram rotates the polarized light by the same angle but in the opposite direction. Such rotation of the plane of polarization is known as **optical activity**. Since the solutions behave as "left" and "right" in the same way as their crystals are left- and right-handed, the optical activity must have its origin in the structure of molecules. LeBel concluded that molecules must also be left- and right-handed, i.e. they must be chiral. Accordingly, the molecules must have spatial structures which are mirror images of one another. Taking into account this evidence, LeBel discovered that organic molecules may be chiral only if the carbon atom is located at the center of tetrahedron. This discovery gave birth to the field of stereochemistry. The same idea about tetrahedral structure and chirality of organic molecules has also been proposed by van't Hoff, who developed his theory independently and using a quite different approach.

Besides the tetrahedral structure, the chirality of a molecule requires one additional condition: all four substituents bound to the central carbon atom must be different. A carbon atom with four different substituents induces chirality of the entire molecule and is called a **stereogenic**, **chiral** or **asymmetric center**. If two or more substituents are identical the molecule and its mirror image can be superimposed and the molecule is not chiral. Some chiral and non-chiral molecules are represented bellow.

Let us examine why only molecules with four different substituents are chiral? What is the fundamental geometric property which the molecule, crystal or hand must possess in order to be chiral? These phenomena can be examined within the **theory of symmetry**. Symmetry is the property which ensures that the figure or geometrical body remains unchanged under particular spatial operation, which is called a **symmetry element**. We say that the object is symmetric in relation to these symmetry elements. Some of the symmetry elements were already described in the chapter on the symmetry of molecular orbitals. Here, let us introduce additional symmetry elements:

SYMMETRY MIRROR, *m*

AXIS of SYMMETRY, C_2

AXIS of SYMMETRY, C_3

(E)-But-2-ene   **(Z)-But-2-ene**

SYMMETRY CENTER   NO SYMMETRY CENTER
OF INVERSION, *i*

The scheme above represents three symmetry elements: **symmetry plane**, **symmetry axis** and the **center of symmetry**. We can recognize that the water molecule, methyl-chloride and formaldehyde all have symmetry planes. The water molecule and formaldehyde also possess one symmetry axis that leaves the molecule unchanged when rotated by 180°. The symmetry axis around which the object can be rotated by 180° angle and remain unchanged is called a **second order**

symmetry axis or **digyre**. The methyl chloride molecule has a symmetry axis with 120° rotation. It is called a **third order axis** or **trigyre**. In the bottom row of the diagram above (*E*)-but-2-ene has a **center of symmetry**, but (*Z*)-but-2-ene does not. A centre of symmetry is the point (labeled with the symbol *i*) through which each atom can be reflected into the equivalent atom with both atoms being at the same distance from *i*. In this way the H-atom in (*E*)-but-2-ene is reflected through *i* into another H-atom. The same principle is valid for methyl groups. The chiral molecules have special symmetry properties because they do not have any symmetry planes as for instance the molecule CHClBrF. Clearly, all molecules in which the carbon atom is bound with the double bond are not chiral and such carbon is not a center of chirality. We know that each C-atom of the double bond is coplanar with three of its neighbors and this plane is at the same time the plane of symmetry. Formaldehyde is a good example. The four different substituents bound to the chiral carbon need not be the atoms of four different chemical elements. Chirality is present if the substituents are different groups or if they have different constitutions. Every chiral molecule with one stereogenic center can have two isomers called **enantiomers**. Enantiomers are related to each other as the object is to its mirror image. Since enantiomers are isomers they can be transformed into each other only by a chemical reaction i.e. by the breaking and forming of chemical bonds. Transformation of one enantiomer into another can involve breaking of any two chemical bonds to the stereogenic centre, switching the positions of the corresponding substituents and the forming of two new chemical bonds.

| ONE ENANTIOMER | REPLACEMENTS OF SUBSTITUENTS (CHEMICAL REACTION) | ANOTHER ENANTIOMER |

If the molecule has more than one stereogenic center, every center can exist in two enantiomeric forms. For **n** stereogenic centers, the number of stereoisomers is 2^n. Stereoisomers of molecules with more than one asymmetric carbon (stereogenic centers) are called **diastereomers**. In constitutional molecular formulas, the stereogenic centers are usually labeled with an asterisks. Some examples of molecules having diastereomers are shown bellow.

In search of chiral C-atoms, it is convenient first to eliminate carbon atoms which cannot be chiral centers. These are for instance CH_2, or CH_3 groups as well as all the C-atoms connected with double or triple bonds.

As already mentioned, the enantiomers rotate the plane of polarized light by a particular angle. Usually, the label (+) is used for the enantiomer whose solution rotates the plane to the right and the label (−) for the one which rotates it to the left. The mixture in which both enantiomers are in equal concentrations does not exhibit optical rotation. Such a solution, known as a **racemate**, does not show rotation of plane polarized light. This is because as one enantiomer rotates to the right by a given angle, the other enantiomer rotates to the left by the same angle so that the rotations cancel each other out. Optical rotation is a characteristic property of a particular compound and is called **specific rotation**, $[\alpha]_D{}^{25}$. Its value is calculated from the measured optical rotation angle α, the length of the measuring container l in dm and the concentration of the sample c in g/100 cm^3. The index D refers to the type of light used in the measurement. In the instruments used for measuring optical rotation, called polarimeters, the source of light is a sodium lamp which emits the specific light called the D-line. Additionally, the superscript 25 indicates that the measurement is made at the 25 °C.

$$[\alpha]_D^{25} = \frac{100[\alpha]}{cl}$$

After LeBel and van't Hoff's discovery that the tetrahedral structure underpins the spatial distribution of atoms in organic molecules, the new concept of **configuration** was introduced. Since the enantiomers differ in their configuration, it was necessary to develop a system of nomenclature by which their configurations can be labeled unequivocally. Cahn, Ingold and Prelog have invented the **system of absolute configuration**, a method similar to the procedure which we have already discussed in the previous chapters. To determine the absolute configuration of a molecule it is necessary to follow certain steps which are similar to the determination of Z and E isomers of alkenes.

1. Determination of the **priorities**.
   The highest priority pertains to the atom with the highest atomic number. In the examples given below, chlorine with the highest atomic number has the priority (1) followed by oxygen (2), carbon (3) and the atom with lowest priority hydrogen (4).

2. Molecular orientation

The molecule must be oriented such a way that the chiral carbon atom is in front and the substituent with the lowest priority (in our example hydrogen) behind this carbon. Frontal atoms and groups are labeled in blue in the next scheme.

3. Determination of the direction of rotation

The molecule is rotated around the axis between the chiral center and the substituent of the lowest priority (see diagram below). The direction of rotation follows the order of priorities of the three substituents. If the rotation has clockwise orientation, the molecule has (**R**) **absolute configuration** and if the rotation is anticlockwise, the molecule will have **absolute configuration** (**S**).

(*R*) CONFIGURATION   (*S*) CONFIGURATION

4. In most organic molecules the atoms directly bound to the chiral carbon are identical. In that case, the rule must be extended so that we look at the atomic number of the next atom along the chain. In the following example we distinguish between the C-atoms in groups CH_3 and CH_3CH_2. In the CH_3 group, the C-atom bound to the stereogenic center is also bound to three hydrogen atoms. In the CH_3CH_2 group the corresponding atom is bound to atoms H, H, C. Therefore since in the CH_3CH_2 group the next atom along the chain is C rather than H, the CH_3CH_2 group has higher priority than CH_3. Hence, the configuration must be (*R*).

5. If the substituent has an atom with a double bond, its priority is considered to be equivalent to two such atoms each bound with the single bond. In the example bellow, the -CHO group has a higher priority than -CH$_2$OH, because the oxygen in -CHO carries a double bond and its priority is equivalent to two oxygen atoms (counted as two C-O single bonds).

Chapter 8
Derivatives of Carboxylic Acids

If the –OH-group is removed from the carboxylic group, the remaining group has the form **RCO-** which is called an **acyl group**. By connecting different atoms or groups to the acyl group we can obtain various structural derivatives of carboxylic acids, like **anhydrides**, **esters**, **acyl halides** or **amides**. The substituent R in the following scheme can be carboxylate RCOO-, alkoxide -OR, halide, for instance -Cl, and amine $-NH_2$, NHR or NR_2.

8.1 Anhydrides

Removing a water molecule from an acid yields an anhydride. Inorganic anhydrides are oxides, so SO_3 is the anhydride of sulfuric acid H_2SO_4. Upon removing water from the carboxylic acids, the molecules condense, yielding an anhydride. If the molecule has only one carboxyl group the condensation involves two molecules, as in the case with acetic acid, where two molecules form **acetanhydride**. Acids with two carboxylic groups, such as **maleic acid** or **phthalic acid**, after the removal of water, form the corresponding cyclic structures.

Acetic acid   **Acetanhydride**

Maleic acid   **Maleic acid anhydride**

Phtalic acid   **Phtalic acid anhydride**

8.2 Esters, Nucleophilic Substitution on the Unsaturated Carbon Atom

An alcohol is a good nucleophile because it has two electron lone pairs on the oxygen atom. In the reaction with a carboxylic acid, the oxygen atom of alcohol attacks the carbon atom of the carboxylic acid group. The mechanism is analogous to the previously discussed nucleophilic additions of aldehydes and ketones. The main difference between the two mechanisms is that the reaction involving carboxylic acid also involves removing a molecule of water. The final product is an **ester**.

Carboxylic acid   **Alcohol**   **Ester**

ESTERIFICATION

The mechanism of this reaction which is called **esterification** is a **nucleophilic substitution on the unsaturated carbon** (the carbon atom is unsaturated because it is bound with the double bond). In summary, esters are formed by the reaction of a carboxylic acid and an alcohol with the elimination of a water molecule. However, removing the water molecule is only possible in an acidic medium, so that in the preparation of esters a few drops of strong acid must be added (see the scheme below). The basic step in this mechanism is the protonation of the oxygen atom of the carboxylic group by a strong acid. This protonated form is susceptible to the nucleophilic attack on the C=O group. The reaction intermediate is transformed into the ester molecule by the removal of a molecule of water:

Ethyl acetate

The nomenclature of esters is relatively simple, because it is similar to the nomenclature of salts. For instance, the ester obtained from methanol and ethanoic acid (acetic acid) is called **methyl ethanoate** (methyl acetate). Esters are compounds that comprise a large number of structures depending on what the alcohol and acid components are. Esters with a small molecular mass are volatile substances, in most cases with a pleasant odor. For instance, butyl acetate is responsible for the odor of apples. In nature, some esters serve as pheromones for insects, for example **isoamyl acetate** which attracts butterflies. Other kinds of esters can be large molecules as for instance waxes and fats where both the alcohol and the acid components can be long-chained or complicated structures. These natural esters will be discussed in the chapter on lipids. In industry, esters are used in the fabrication of polymeric fibers for textile materials and plastic materials for various uses. The most common material is **polyethylene terephthalate (PET)**, the ester prepared from terephthalic acid and ethylene glycol.

| **Ethanoic acid** (Acetic acid) | **Butanol** | **Butyl ethanoate** (Butyl acetate) |

| **Ethanoic acid** (Acetic acid) | **Isoamyl alcohol** | **Isoamyl acetate** |

HOOC—⬡—COOH   HOOC—⬡—COOH   HOOC—⬡—COOH
         HO   OH   HO   OH
Terephthalic acid
       Ethylene glycol

                              │ - *n* H₂O
                              ↓

Polyethylene glycol terephthalate
(PET)

One of the most typical reactions of esters is the **saponification** or **hydrolysis of esters** by strong bases such as NaOH or KOH. Boiling ester with a water solution of KOH or NaOH yields the salt of a carboxylic acid and the corresponding alcohol.

Methyl benzoate   →NaOH→   Sodium benzoate   + CH₃OH   Methanol

Salts can simply be transformed into a carboxylic acid just by reacting with a strong acid. Remember the principle that strong acids can yield weak acids from their salts:

Sodium benzoate   →HCl→   Benzioc acid   + NaCl

Analogously with aldehydes and ketones, esters can be reduced by strong reducing agents such as LiAlH₄. Weaker reducing reagents such as NaBH₄ can reduce only aldehydes and ketones, but are unreactive towards esters. The reaction with LiAlH₄ must be performed in dry ether. The obtained intermediate cannot be isolated but is immediately hydrolyzed with diluted acid. The products are two alcohols, the first from the acid component and the second from the alcohol component of the ester precursor.

CH₃CH₂—C(Ö:)(OCH₂CH₃)   →1) LiAlH₄ 2) H⁺, H₂O Diethyl ether→   CH₃CH₂CH₂OH + CH₃CH₂OH
Ethyl propanoate   Propanol   Ethanol

8.3   Acyl Halides

Replacing the OH group in a carboxylic acid with a halogen generates acyl halides, the reactive compounds which are important intermediates in the organic synthesis of various other compounds, especially amides. The most frequently used acyl halide is acyl chloride, the compound prepared from carboxylic acid and a special reagent called **thionyl chloride** $SOCl_2$. This reagent has an unpleasant odor because it readily reacts with water, yielding HCl and SO_2.

$$SOCL_2 + H_2O \rightarrow SO_2 + 2HCl$$

Thionyl chloride reacts with carboxylic acids in the way as shown in the scheme below. In the first step the reaction yields HCl and the intermediate benzoyl chlorosulfonate:

| Benzoic acid | Thionyl chloride | Benzoyl chlorosulfate |

Benzoyl chlorosulfonate immediately rearranges into the acyl chloride (benzoyl chloride in our example) and SO_2. This reaction is an interesting example of internal nucleophilic substitution. The internal nucleophile is the chlorine atom (it is electron-rich) which attacks the carbonyl carbon. The final products are benzoyl chloride and SO_2.

Benzoyl chloride

The nomenclature of acyl halides is simple. Their names are deduced from the name of the constituent acid with the suffix **-yl** or **-oyl** and the name of the halide.

Methanoyl chloride
(Formyl chloride)

Ethanoyl bromide
(Acetyl bromide)

Propanoyl chloride

Like thionyl chloride, acyl halides are sensitive to moisture and quickly hydrolyze into the corresponding carboxylic acid and HCl. This is an example of the nucleophilic substitution at the unsaturated carbon with H_2O as the nucleophile.

Propanoyl chloride   **Propanoic acid**

Besides OH⁻ or water, other nucleophiles can also be involved in this type of reaction. If a molecule of alcohol is the nucleophile, the products are ethers. The reaction is faster if HCl is removed immediately from the reaction mixture. In general, hydrogen halides can be removed efficiently from the reaction mixtures by the addition of an amine which, together with hydrogen halide, forms the corresponding ammonium salt (this has already been discussed in previous chapters). In the following example the amine used is pyridine.

Ethanoyl chloride   **Ethyl ethanoaet**
(Acetyl chloride)   **(Ehyl acetate)**

Pyiridine

8.4 Amides

Perhaps the most important synthetic use of acyl halides is in the reactions with ammonia or with amines. The products are important natural compounds called **amides**. Depending on the structure of the nucleophiles, i.e. whether they are ammonia, primary or secondary amines, the products are **primary**, **secondary** or **tertiary amides**.

Methanoyl chloride   **Methylamide**
(Formyl chloride)   **(Formamide)**

PRIMARY AMIDE   SECONDARY AMIDE   TERTIARY AMIDE

$$HC\underset{NH_2}{\overset{O}{\diagup}}\qquad HC\underset{\underset{H}{\overset{|}{N}\cdot CH_3}}{\overset{O}{\diagup}}\qquad HC\underset{\underset{CH_3}{\overset{|}{N}\cdot CH_3}}{\overset{O}{\diagup}}$$

Methanamide   *N*-**Methylmethanamide**   *N*,*N*-**Dimethylmethanamide**
(**Formamide**)   (*N*-**Methylformamide**)   (*N*,*N*-**Dimethylformamide**)

In amides, the amino group is directly bonded to the carbonyl group. Amide nomenclature includes the simple suffix **-amide**. In secondary and tertiary amides, the substituents that are bound to the nitrogen atom are not labeled by the number, but by the letter *N*. The representative examples are *N*-methylmethanamide or *N*,*N*- dimethylmethanamide in the above scheme. Some amides which are derivatives of formic or of acetic acids also bear traditional names as **formamide** or **acetamide**.

Since amides have two different heteroatoms with lone pairs, the electrons can be delocalized over both N- and O-atoms. Let us discuss the electron delocalization by using the method of resonance structures:

$$\left[\quad RC\underset{NH_2}{\overset{O:}{\diagup}}\quad \longleftrightarrow \quad RC\underset{\overset{+}{N}H_2}{\overset{\overset{-}{O}:}{\diagup}}\quad\right]$$

Two important conclusions can be drawn from the given resonance structures. First, the electron lone pair is not localized on the nitrogen atom as in saturated amines. We recall from the previous chapter that amines are less basic if the nitrogen lone pair is not localized on the N-atom. This was illustrated by an example where we compared the basicities of aniline and pyridine and showed that aniline is a weaker base because its lone pair electrons are delocalized over the entire benzene ring. For the same reason, **amides** are neutral rather than basic.

The second very important conclusion is related to the nature of the covalent bond between the nitrogen and the carbon atom. This bond has a single bond character in one of the resonance structures and a double bond character in the other. Hence, the amide bond has a bond order of 1.5, i.e. it has an intermediate character between a double and a single bond. In the chapter on alkanes we have mentioned that while the rotation around a single bond is free, the rotation around a double bond is forbidden. Because the amides bond has a partial double bond character the rotation around it is hindered. This evidence is important in the study of the stereochemistry of complex molecules with amide groups, such as polyamides and polypeptides.

One of the best known polyamides is **nylon**, the substance which was first produced in 1935. Nylon can be obtained from hexamethylenediamine and adipic acid by a procedure called **polycondensation**.

$$n\,H_2N(CH_2)_6NH_2 \;+\; n\,HOOC(CH_2)_4COOH \longrightarrow \left[NH(CH_2)_6NH\overset{\displaystyle O}{\overset{\|}{C}}(CH_2)_4C\right]_n$$

1,6-Diaminohexane   **Adipinic acid**   **Nylon**
(Hexamethylene diamine)   **(Polyamide)**

Many antibiotics and other pharmaceutical products have molecules with amide groups. A large group of such compounds are antibiotics derived from penicillin (see the scheme below). These compounds are stable because the amide bond is difficult to hydrolyze.

Penicillin G

(antibiotic)

Chapter 9
Electrophilic Substitutions

Reactions of Aromatic Compounds

In the chapter about hydrocarbons we have discussed aromatic compounds, their stability and specific chemical behavior. The electrons in π-orbitals are delocalized around the benzene ring and although there are formal double bonds, aromatic compounds do not undergo standard addition reactions typical for alkenes. The distribution of π-electrons in the ring is such that the highest electron density is above and below the plane of the aromatic ring. Such π-electron distribution is the reason that the benzene ring behaves like a Lewis base, which can donate electrons or react with strong Lewis acids. Protons and other positively charged particles are **electrophiles** because they tend to bind to species which are electron rich. Consequently, benzene and other aromatic molecules can be protonated by strong acids such as H_2SO_4.

The cation which is formed by protonation is stabilized because of the delocalization of electrons, as shown in the next scheme. For simplicity, these resonance structures can be replaced with a single formula where the region in which the electrons are delocalized is designated with dotted line and the + sign in the middle.

H. Vančik, *Basic Organic Chemistry for the Life Sciences*,
DOI 10.1007/978-3-319-07605-8_9, © Springer International Publishing Switzerland 2014

In general, the electrophile is labeled with E^+ and its addition to the benzene ring is analogous to protonation (see scheme below). If the cationic reaction intermediate loses a proton, the electrophile remains attached to the ring. The final product has the structure in which hydrogen is replaced with the incoming electrophile and such a reaction is called an **electrophilic substitution** on the unsaturated carbon atom.

E+ = ELECTROPHILE

If we wish to replace one hydrogen atom with some functional group or atom, this group or atom must be transformed into the cationic, electrophilic form. For instance, for the halogenation of a benzene ring, the halogen atom must appear in its electrophilic form called a **halonium ion**. Since such ions are unstable in solution so they must be prepared *in situ* (immediately, without isolation) and in the form of the complex with a larger molecule. Such reactants, with a positive halogen, appear in the reaction of halogens with metal halides such as $FeCl_3$, $FeBr_3$, or $AlCl_3$.

$$Br_2 + FeBr_3 \longrightarrow Br^+[FeBr_4]^- \quad \text{or} \quad Cl_2 + AlCl_3 \longrightarrow Cl^+[AlCl_4]^-$$

Bromonium or **chloronium** complexes, which are shown in the examples above, are good electrophiles and react with benzene by replacing the hydrogen atom with the halogen.

Brombenzene

Alkyl or acyl groups can be bound to benzene in an analogous way. These reactions are called **Friedel-Crafts alkylations** or **acylations**, respectively, in honor of their discoverers **Charles Friedel** and **James M. Crafts**. The starting compound in alkylation is an alkyl halide which, together with the corresponding metal halide, forms a molecular complex that consists of a carbocation and a tetrahalometallic anion. For instance, ethyl bromide reacts with $FeBr_3$ forming a complex in which the ethyl group has a cationic character and behaves as an electrophile. This electrophile reacts with benzene molecule as shown in the following scheme.

$$CH_3CH_2Br \ + \ FeBr_3 \ \longrightarrow \ CH_3\overset{+}{C}H_2 \left[FeBr_4\right]^{-}$$

Ethylbenzene

In acylation the starting compound is the corresponding acyl-chloride:

Phenylmethyl ketone

Reaction of nitric acid with sulfuric acid yields the **nitronium ion** (NO_2^+), which as an electrophile reacts with benzene and forms **nitrobenzene**.

$$HNO_3 \ + \ 2\,H_2SO_4 \ \longrightarrow \ \overset{+}{N}O_2 \ + \ H_3\overset{+}{O} \ + \ 2\,HSO_4$$

Nitronium ion

Nitrobenzene

9.1 Substituent Effects in Electrophilic Aromatic Substitution

The chemistry of substituted benzenes belongs to the field of organic synthesis which was developed during the nineteenth century for the needs of the dye industry. Aromatic compounds which have several substituents on the benzene ring have been of special practical interest. The problem of introducing the second substituent arises because this additional group can bind at different positions along the ring.

From their long experience in synthesis the chemists have discovered that the first substituent on benzene plays a crucial role in influencing (**directing**) the position of the second substituent group. Some functional groups direct the next substituent into the *ortho* or *para* position while others favor a *meta* substitution. The groups which direct into the *ortho* or *para* position are called **activating** substituents while those which direct into the *meta* position are called **deactivating** groups. In principle, the term activating implies the donation of electrons to the benzene ring and deactivating has the opposite effect, i.e. the withdrawal of electrons from the ring.

Experience has shown that substituents with electron lone pairs on the atom directly bound to benzene are activating and they are also ***ortho*** or ***para*-directors**, i.e. they guide the second substituent into *ortho/para* positions. The deactivating substituents do not have such atoms, but they can have a double bond or an electronegative atom. The deactivating substituents are ***meta*-directors**, i.e. they display regioselectivity by directing the second substituent into the *meta* position. The alkyl groups are a special case, they are activating in spite of their lack of a heteroatom with an electron lone pair.

ACTIVATING GROUPS
R_{ACT}

Aniline Cl_2, $AlCl_3$ o-Chloroaniline   +   p-Chloroaniline

DEACTIVATING GROUPS
R_{DE}

Nitrobenzene Cl_2, $AlCl_3$ m-Chloronitrobenzene

ACTIVATING GROUPS R_{ACT}	DEACTIVATING GROUPS R_{DE}
$-\ddot{N}H_2$	$-NO_2$
$-\ddot{N}HR$	$-CF_3$
$-\ddot{N}R_2$	$-SO_3H$
$-\ddot{O}H$	$-CN$
$-\ddot{O}R$	$-\overset{O}{\overset{\|}{C}}Cl$
$-\ddot{X}:$ for X = Cl, Br, I	$-\overset{O}{\overset{\|}{C}}OH$
$-R$ for R = alkyl	

These experimental observations have played a crucial role, not only in the development of the theory of reaction mechanisms, but also in the final solution of the problem of the electronic structure of benzene. Historically, the explanation of the laws of substituent direction appeared at two levels. During the first quarter of the last century, when the theory of resonance was in the early stages of development, substituent directing was explained by the charge distribution in the reactant molecule. As we can see in the example of the aniline molecule, the activating $-NH_2$ group causes the negative charge to appear only in the *ortho* or *para* positions. Since they are negatively charged, these positions are preferred for an electrophilic attack.

Conversely, a deactivating group such as $-NO_2$ in nitrobenzene induces a positive charge in the *ortho* and *para* positions, so the electrophilic attack will not take place at these positions. Instead, the electrophile will attack the *meta* position.

These initial ideas were inadequate to explain the orientation effect of all groups. The problem appeared with the methyl group in toluene, which is activating, but for which no resonance structures could be constructed. The proper explanation of the regioselectivity (the preference for certain substituent positions) only came with the development of the theory of reaction mechanisms by Robinson and Ingold. As we have seen at the beginning of this chapter, the electrophilic attack causes the formation of a reaction intermediate, the carbocation. Since we know that the rate of chemical reaction depends mostly on the stability of the reaction intermediate, let us discuss the structure of the activated and deactivated cations.

In the following example, the activating substituent is the methoxy group. If the electrophile attacks the *ortho* position the corresponding cation can be described by the resonance structures shown in the following scheme.

OXONIUM ION

The electron-donating property of the methoxy group is most evident in the last resonance structure, describing the **oxonium ion**, with a positive charge on the oxygen. The cation is stabilized by electrons transferred from the oxygen lone pairs. This is why all groups with electron lone pairs on the atom directly bound to the benzene ring are activating. Analogous resonance structures can also be written for aniline.

A methyl group, which does not have electron lone pairs, is activating because it forms *inter alia* a stable tertiary carbocation (scheme below).

TERTIARY
CATION

An electrophilic attack on *para* position leads to the formation of an inter-mediate the resonance structure of one of these is also stable tertiary carbocation. In conclusion, the activating substituents direct the next group into both the *ortho* and *para* positions.

OXONIUM ION

The electrophilic attack on *meta* position would yield a carbocation that cannot be stabilized by the activating group, so this position is not favored. This can clearly be seen from the resonance structures in the next scheme. The carbon on which the activating group is bound always has a double bond and the delocalization of electrons from the methoxy group into the ring is not possible. The corresponding carbocation is not stabilized and activating groups do not support the formation of *meta*-products.

The deactivating groups withdraw electrons from the ring and the binding of an electrophile will yield an intermediate in which the resonance structures can be written as shown in the scheme below. In the *ortho* or *para* position, the electrophile induces the formation of intermediates in which two positive charges are located next to each other (labeled with dashed circle in the scheme), which is energetically unfavorable. However, this situation is avoided if the electrophile is bound in the *meta* position. Hence the deactivating group is always a *meta*-director.

ortho

UNFAVORED

para

UNFAVORED

meta

FAVORED

The deactivating effect of the CF$_3$ group is the consequence of a strong electron withdrawing of electron density from the benzene ring via the inductive effect of three fluorine atoms.

A typical example of the substituent-director effect (regioselectivity) is the preparation of the well-known explosive **trinitrotoluene (TNT)** by nitration of toluene. A methyl group is activating and directs the nitro substituents into the *ortho* and *para* positions.

Chapter 10
Cycloadditions

In the first chapters of this book we have mentioned polycyclic compounds. Reactions in which polycyclic compounds are formed have been known for more than a hundred years, but their mechanisms remained puzzling for a long period of time. These reactions were even called "reactions without a mechanism". The best known of this type of reactions, is the **Diels-Alder addition**, named in the honor of **Otto Diels** and **Kurt Alder** in which two alkene molecules are condensed so that two new σ-bonds are formed. Good examples are reactions in which the norbornene skeletons are obtained.

The occurrence of these reactions strongly depends on the electronic structures of the reactants, especially on the number of π-electrons. It has been shown that the reactions are favored if one of the reactants has four π-electrons and the other has two π-electrons (4 π + 2 π addition). These six π electrons are rearranged in the product into two new σ bonds (labeled in red in the scheme) and one new π bond.

4 π Electrons   2 π Electrons

Other combinations of reactants such as two **π** electrons with two **π** electrons or four **π** electrons with four **π** electrons do not react under thermal conditions, but readily react if they are irradiated with visible or ultraviolet light. These rules of reactivity in cycloadditions were discovered by **Robert B. Woodward** and **Roald Hoffmann** on the basis of quantum-mechanical analysis and the symmetry properties of molecular orbitals. These rules are known as the **Woodward-Hoffmann rules**.

4 π Electrons   4 π Electrons

2 π Electrons   2 π Electrons

Chapter 11
Organic Natural Products

The knowledge of basic structural and dynamic concepts of organic chemistry allows us to examine some aspects of the chemistry of living organisms. The branch of organic chemistry that deals with compounds isolated from the living organisms is called **chemistry of natural products**. Basic knowledge of this field of chemistry is of fundamental importance for the understanding of biochemistry.

In principle, natural products can be divided in two categories, **primary** and **secondary metabolites**. Primary metabolites are compounds that are necessary for the functioning of the organism and with minor variations they are common to all living organisms. This group comprises amino acids, carbohydrates, lipids and nucleotides. Some organic compounds that are characteristic of a specific biological genus belong to secondary metabolites. Good examples of secondary metabolites are alkaloids, some terpenes, flavonoids, etc. These classes of compounds are not common to all organisms. The literature definition states that the **chemistry of natural products** has its main interest in secondary metabolites while the science interested in primary metabolites is called **bioorganic chemistry**. Bioorganic chemistry is closely related to **biochemistry**, which investigates the chemical systems in living organisms.

Our discussions about the natural products will begin with the primary metabolites. Since the organic compounds in nature always exist in complex mixtures, chemists developed special laboratory methods for their purification and isolation. One of the earliest of such methods is extraction, the technique that is based on the different solubility of substances. Because similar substances dissolve in similar solvents, most organic substances will be soluble in organic solvents: polar substances in polar solvents, nonpolar substances in nonpolar solvents. In contrast, ionic compounds are well soluble in water. Amongst primary metabolites, the most water soluble compounds are amino acids and polar carbohydrates. A large group of structurally diverse natural products soluble in nonpolar solvents are lipids which will be discussed at the end of this chapter.

H. Vančik, *Basic Organic Chemistry for the Life Sciences*,
DOI 10.1007/978-3-319-07605-8_11, © Springer International Publishing Switzerland 2014

11.1   Amino Acids and Peptides

The physiologically most important components of living beings are proteins, the carriers of basic physiological and biochemical functions. Their molecules are extraordinarily complex and varied but in spite of this complexity their structures are built from only 20 relatively simple molecules – the **amino acids**. Before investigating these biologically important amino acids which are the products of hydrolysis of proteins, we shall describe the structure and properties of this class of compounds in general.

The molecules of amino acids have two functional groups with different properties, the basic amino group and the acidic carboxyl group. Because the amino group can be bound to different carbon atoms on the hydrocarbon chain, the names of these compounds are derived from the position of this functional group. However, for this class of compounds, the traditional nomenclature in which the C-atoms are labeled not by numbers, but by letters of the Greek alphabet is still in use. In addition, the letter α does not correspond to the carbon atom labeled with number 1, but to the atom with the number 2. In this nomenclature there are **α, β, γ amino acids**, etc.

$$\begin{array}{ccc} \gamma & \beta & \alpha \\ 4 & 3 & 2 \quad 1 \end{array}$$
CH$_3$CH$_2$CHCOOH
|
NH$_2$

2-Aminobutanoic acid

α-Aminobutanoic acid

$$\begin{array}{ccc} \gamma & \beta & \alpha \\ 4 & 3 & 2 \quad 1 \end{array}$$
CH$_3$CHCH$_2$COOH
|
NH$_2$

3-Aminobutanoic acid

β-Aminobutanoic acid

$$\begin{array}{ccc} \gamma & \beta & \alpha \\ 4 & 3 & 2 \quad 1 \end{array}$$
H$_2$NCH$_2$CH$_2$CH$_2$COOH

4-Aminobutanoic acid

γ-Aminobutanoic acid

All 20 amino acids obtained from natural proteins belong to the **α-amino acids** and can be represented by the general formula H$_2$NCHRCOOH. They differ only in the substituent R.

Amino acids are amphoteric compounds; the carboxylic groups given them acidic properties and the amino groups give them a basic character. Their amphoteric behavior is shown in the next scheme.

AMINO ACID AS ACID

AMINO ACID AS BASE

Amino acids are well soluble in water, because they can appear in the form of a dipolar ion called, by the German word, Zwitterion. This ion can be considered as the product of self-protonation: the amino group is protonated by the proton that comes from the carboxyl group.

$$\overset{\displaystyle ..}{H_2N}CHCOOH \underset{-H^+}{\overset{+H^+}{\longrightarrow}} \overset{+}{H_3N}CHCOO^- \qquad \text{DIPOLAR ION}$$
$$\qquad\qquad | \qquad\qquad\qquad\qquad | \qquad\qquad\qquad \text{ZWITTERION}$$
$$\qquad\qquad R \qquad\qquad\qquad\qquad R$$

When the R group is neutral the pH of the solution of such an amino acid would be approximately 7, i.e. this amino acid is neutral. However, R groups can be neutral, acidic or basic. For instance, in aspartic acid (see following scheme) the group R has a carboxylic functional group. Consequently, pH of a water solution of aspartic acid must be lower than 7 and it is actually 2.95. On the other hand, in the amino acid lysine, the R group contains an amino group and the compound behaves as a base. In water solution, lysine has a pH equal to 9.8. The pH value of the amino acid in water solution depends on the functional group R and is called the **isoelectric point**, which is labeled with the symbol **pI**.

$$\overset{H}{\overset{|}{^-OOCCH_2CCOO^-}} \text{ Aspartic acid} \qquad \overset{H}{\overset{|}{H_3^+NCH_2CH_2CH_2CCOO^-}} \text{ Lysine}$$
$$\qquad \underset{+}{NH_3} \qquad \mathbf{pI = 2,95} \qquad\qquad\qquad \underset{+}{NH_3} \qquad \mathbf{pI = 9,8}$$

In the following schemes all 20 amino acids are given and classified into neutral, basic and acidic forms together with their pI values. Since in biochemistry we handle complex protein molecules consisting of thousands of amino acids, the writing of exact formulas becomes unpractical, so every amino acid is labeled by three letters. All three-letter labels are listed in the schemes that follow.

BAISC AMINO ACIDS

$$\overset{H}{\overset{|}{H_3^+NCH_2CH_2CH_2CH_2CCOO^-}} \qquad\qquad \text{Lysine} \qquad\qquad \mathbf{Lys}$$
$$\qquad\quad \underset{+}{NH_3} \qquad\qquad\qquad\qquad \mathbf{pI = 9,8}$$

$$\overset{+}{\overset{NH_2}{\overset{\|}{H_2NCNHCH_2CH_2CH_2CCOO^-}}} \qquad \text{Arginine} \qquad \mathbf{Arg}$$
$$\qquad\qquad\qquad \underset{+}{NH_3} \qquad\qquad \mathbf{pI = 10,75}$$

$$\text{(imidazole)} -CH_2\overset{H}{\overset{|}{C}}COO^- \qquad \text{Histidine} \qquad \mathbf{His}$$
$$\qquad\qquad\qquad \underset{+}{NH_3} \qquad\qquad \mathbf{pI = 7,65}$$

ACIDIC AMINO ACIDS

$$\overset{\displaystyle H}{\underset{\displaystyle \overset{+}{NH_3}}{{}^-OOCCH_2\overset{|}{\underset{|}{C}}COO^-}}$$

Aspartic acid

pI = 2,95

Asp

$$\overset{\displaystyle H}{\underset{\displaystyle \overset{+}{NH_3}}{{}^-OOCCH_2CH_2\overset{|}{\underset{|}{C}}COO^-}}$$

Glutamic acid

pI = 3,1

Glu

NEUTRAL AMINO ACIDS WITH POLAR GROUPS R

$$\underset{\displaystyle \overset{+}{NH_3}}{\overset{\displaystyle O \quad H}{H_2N\overset{||}{C}CH_2\overset{|}{\underset{|}{C}}COO^-}}$$

Asparagine

pI = 5,40

Asn

$$\underset{\displaystyle \overset{+}{NH_3}}{\overset{\displaystyle O \quad\quad H}{H_2N\overset{||}{C}CH_2CH_2\overset{|}{\underset{|}{C}}COO^-}}$$

Glutamine

pI = 5,65

Gln

$$\overset{\displaystyle H}{\underset{\displaystyle \overset{+}{NH_3}}{HOCH_2\overset{|}{\underset{|}{C}}COO^-}}$$

Serine

pI = 5,70

Ser

$$\underset{\displaystyle \overset{+}{NH_3}}{\overset{\displaystyle OH\,H}{CH_3\overset{|}{C}H\overset{|}{\underset{|}{C}}COO^-}}$$

Threonine

pI = 5,60

Thr

$$HO\!-\!\!\bigcirc\!\!-\!CH_2\underset{\displaystyle \overset{+}{NH_3}}{\overset{\displaystyle H}{\overset{|}{\underset{|}{C}}COO^-}}$$

Tyrozine

pI = 5,70

Tyr

$$\overset{\displaystyle H}{\underset{\displaystyle \overset{+}{NH_3}}{HSCH_2\overset{|}{\underset{|}{C}}COO^-}}$$

Cysteine

pI = 5,15

Cys

NONPOLAR AMINO ACIDS WITH NONPOLAR SIDE GROUPS R

Structure	Name	Abbreviation
$\overset{\displaystyle H}{\underset{\underset{+}{NH_3}}{HCCOO^-}}$	Glycine pI = 10,75	**Gly**
$\overset{\displaystyle H}{\underset{\underset{+}{NH_3}}{CH_3CCOO^-}}$	Alanine pI = 6,15	**Ala**
$\overset{\displaystyle H}{\underset{\underset{+}{NH_3}}{(CH_3)_2CHCCOO^-}}$	Valine pI = 6,00	**Val**
$\overset{\displaystyle H}{\underset{\underset{+}{NH_3}}{(CH_3)_2CHCH_2CCOO^-}}$	Leucine pI = 6,00	**Leu**
$\overset{CH_3\ \ H}{\underset{\underset{+}{NH_3}}{CH_3CH_2CH-CCOO^-}}$	Isoleucine pI = 6,05	**Ile**
$\overset{\displaystyle H}{\underset{\underset{+}{NH_3}}{CH_3SCH_2CH_2CCOO^-}}$	Methionin pI = 5,70	**Met**
Proline structure	Proline pI = 6,30	**Pro**
$\overset{\displaystyle H}{\underset{\underset{+}{NH_3}}{C_6H_5-CH_2CCOO^-}}$	Phenylalanine pI = 5,75	**Phe**
$\overset{\displaystyle H}{\underset{\underset{+}{NH_3}}{\text{indole}-CH_2CCOO^-}}$	Tryptophan pI = 5,95	**Trp**

The amino acids valine, leucine, isoleucine, methionine, phenylalanine, tryptophane, threonine, lysine, arginine and histidine are called **essential**, because they are necessary components of the food of mammals. Namely, mammalian organisms are not able to synthesize amino acids in this group.

In all the listed amino acids, with the exception of glycine, the α-carbon is bound
to four different substituents hence it is a stereogenic center. From this it follows
that every amino acid can appear in the form of two enantiomers. In the following
example, both the enantiomers of alanine are represented together with their absolute configurations. However, enantiomers of amino acids can also be represented
by the traditional notation of chiral molecules that is called the **relative configuration**. This nomenclature for configuration was proposed **Emil Fischer** in the nineteenth century for the representation of the stereochemistry of carbohydrates.

For the determination of the relative configuration, the molecule is oriented so
that the chiral carbon atom is in the center of the cross, the substituents positioned
at the top and bottom arms of the cross lie bellow the plane of drawing and the
groups at the left and right arm are above the plane of the drawing. Substituents H
and NH_2 are always either the left or the right arm and the carboxylic group is on the
top arm of the cross. These representations are called **Fischer formulas**. If the NH_2
group is on the left arm of the cross, the configuration is designated as the
L-configuration. Conversely, if NH_2 is on the right arm the molecule has the
D-configuration. Relative and absolute configurations of alanine are presented in
the next scheme. It must be pointed out that absolute and relative configurations do
not correlate: (R) is not always D and (S) is not always L.

It is important to mention that **all 20 amino acids obtained from living organisms have the same L relative configuration**. Although science does not have the
full explanation for this observation, such stereoselectivity supports the notion that
chemical and later biological evolution started from the same enantiomer.

FISCHER`S FORMULAS

The laboratory synthesis of amino acids is fundamentally different from the way
these compounds are synthesized in the living organisms (**biosynthesis**). The traditional synthesis of amino acids is known as **Strecker synthesis** (in the honor of
Adolph Strecker). This method is based on the nucleophilic attack on the carbonyl
group; a reaction which we already discussed in this book. In this reaction of aldehyde with ammonium chloride and sodium cyanide the product is aminonitrile. By
boiling with strong acid, this aminonitrile is transformed into the corresponding
amino acid.

Acetaldehyde   2-Aminopropanenitrile   L-Alanine + D-Alanine
    RACEMATE

From the chemical equations above it is clear that the laboratory synthesis is not a "clean" reaction, because it yields a mixture of the both enantiomers: L-alanine and D-alanin i.e. a racemate. The appearance of the racemate can be explained from the knowledge of the reaction mechanism of nucleophilic addition to the carbonyl group. In the carbonyl group the carbon atom is bound by a double bond and lies with all three of its substituents in the same molecular plane. The CN^- group as a nucleophile can attack the carbonyl carbon with the same probability from either side of the molecular plane. Consequently above-the-plane attack yields (R) enantiomer and below-the-plane attack gives the (S) enantiomer.

The biosynthesis of amino acids in living organisms is a more complex process that involves biocatalysts, enzymes. This biocatalytic process is stereoselective, meaning that it leads to only one enantiomer as a product, the enantiomer having L-configuration.

One of the known mechanisms of biosynthesis is shown in the next scheme. The enzyme **aminotransferase** transfers the amino group from the precursor amino acid to the **α-ketoacid**, which is rearranged into the product amino acid. In this process, the precursor amino acid is transformed into the corresponding **α-ketoacid**. It must be pointed out that besides transferring the amino group, the enzyme aminotransferase also preserves the stereochemistry: the L-precursor amino acid leads to the formation of the L-product aminoacid. This mechanism presented below is for the biosynthesis of L-glutamic acid.

α-Ketoglutamic acid

L-Glutamic acid

enzyme
AMINOTRANSFERASE

L-Amino acid

α-Ketoacid

The most important property of amino acids is their ability to form dimers, oligo-mers and polymers that are called **peptides** and **polypeptides**. The mechanism of the formation of peptides and polypeptides is complex and will not be discussed in this book. Let us only mention that the **peptide bond** is formed by connecting the amino group of one amino acid to the carboxyl group of the other amino acid followed by the elimination of water.

PEPTIDE BOND

Cys

Cysteine

Ser

Serine

Cys - Ser

Cysteylserine

DIPEPTIDE

By convention, the peptides are represented by connecting the short labels of amino acids in such a way that the first amino acid on the left hand side has a free amino group and the last amino acid on the right hand side ends with the carboxyl group. By using this convention, the dipeptide cysteyl-serine was represented in the scheme above. In reality, the **peptide bond** is the **amide group** because there is an amino group directly bonded to the carbonyl group. Large numbers of amino acids can be connected into polypeptides that are represented in the same way as dipeptides. The first and the last amino acid in the polypeptide chain are called **terminal amino acids**. Accordingly, the left and the right end of the chain are named the **N-terminus**, and **C-terminus**, respectively.

Longer polypeptide chains can be interconnected by the formation of the **disulfide bond** as shown for the two polypeptides in the scheme below.

In the same way the long polypeptide chain can also form a loop as shown below:

We have argued that the peptide bond is essentially related to the amide functional group. The resonance structures of amides, which we have discussed earlier, can also be used in the representation of the electronic structure of polypeptide molecules.

Inspection of these structures makes it clear that the chemical bonds between carbon and nitrogen have a double bond character. Consequently, the rotation around these bonds is restricted as is the rotation around the double bonds in alkenes. This restricted rotation makes polypeptide chains rigid. Because of structural rigidity polypeptide chains appear in two forms: the **α-helix** and the **β-sheet**.

Each complete turn of the helix requires 3.6 amino acids (residues) in the chain which contains N-C-C-N bonds. The helix is additionally stabilized by hydrogen bonds between the CO group of residue n and the NH group of residue n+4 i.e. between spatially close amino acids (see next diagram).

0, 56 nm

3,6 nm distance
from N to C terminus

AXIS OF THE
α - HELIX

In β-sheets the chains are oriented with respect to each other in such a way that the N-H groups in the backbone of one strand establish hydrogen bonds with the C=O groups in the backbone of the adjacent strand. The chains can be interconnected in two ways: **antiparallel** or **parallel**. In the antiparallel arrangement, as represented in the scheme below, the successive β strands alternate directions so that the N-terminus of one strand is adjacent to the C-terminus of the other strand.

From N- to C-terminus

HYDROGEN
BONDS

From C- to N-terminus

β - SHEET

In the parallel structure shown below the N-termini in both strands are oriented in the same direction as are the C-termini.

From N- to C-terminus

HYDROGEN
BONDS

From N- to C-terminus

β - SHEET

Since the **protein molecules** consist of polypeptide chains interconnected by disulfide bonds, their structure can be represented by simple graphical models in which the α-helices and β-sheets are depicted as in the figure below.

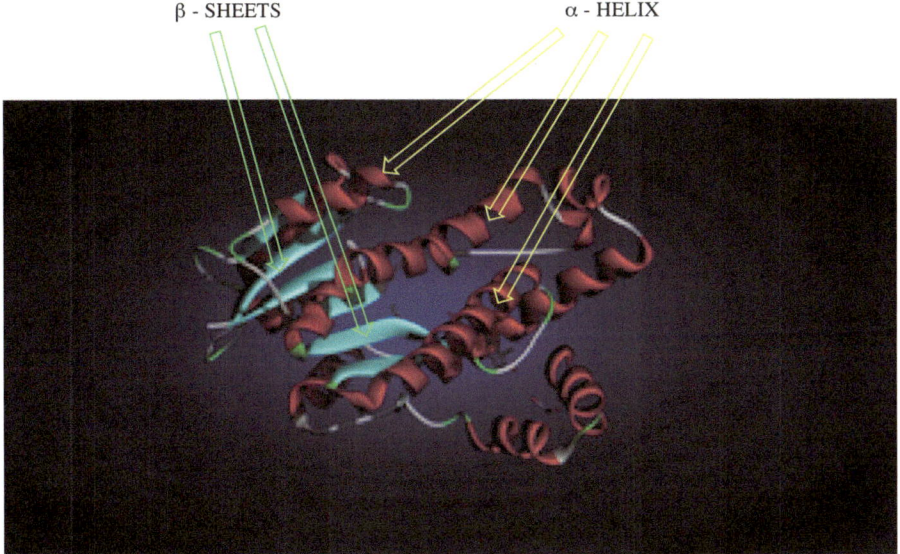

As we have seen from the overview of chemistry of polypeptides, three levels of the structure of proteins can be discerned. The first level is the **primary structure** of proteins (polypeptides) which is simply the linear order of amino acids in the chain starting from the N-terminus up to the C-terminus. Formation of α-helices and β-sheets is the **secondary structure of proteins** and it is responsible for the shape of the polypeptide molecule. Aggregation of these polypeptides into the final molecule of the protein is its **tertiary structure** (as shown in the figure above). The most common way by which polypeptides are interconnected into the tertiary structure includes the disulfide bonds between the cysteine residues (amino acids).

11.2 Carbohydrates

Another class of primary metabolites soluble in water are the carbohydrates. To this class belong the substances which we know as sugars, but also polymers such as starch and cellulose, compounds that are ubiquitous components of the living organisms. The first discovery in the chemistry of sugars was the observation that some of them yield other carbohydrates by hydrolysis. In contrast, there are also sugars that cannot be further hydrolyzed. They are called **monosaccharides**. The molecules of monosaccharides can be joined into the larger molecules called **oligosaccharides** or even **polysaccharides**.

The name "carbohydrates" comes from their composition which is represented by the general formula $C_m(H_2O)_n$ which resembles the combination of carbon and water. This composition implies that a molecule of monosaccharide consists of a hydrocarbon chain with attached hydroxyl groups. More detailed analysis shows that most of the monosaccharide molecules have hydrocarbon chains which are from five to seven carbon atoms long. From the study of the chemical behavior of different monosaccharides it follows that some of them exhibit reactions typical for aldehydes and others show reactions typical for ketones. The monosaccharides can therefore be classified into two groups: **aldoses** and **ketoses**. In their nomenclature, the names of carbohydrates are formed from the root based on the number of C-atoms, the fundamental functional group (ketone or aldehyde) and the suffix **-ose**. While the aldehyde group always contains the first carbon atom of the chain, keto-group in all known carbohydrates appears at the second carbon atom of the chain. Some monosaccharides named in accordance with these rules are represented in the scheme below.

H O	H O	1 CH₂OH	1 CH₂OH
1 C	1 C		
2 CHOH*	2 CHOH*	2 C=O	2 C=O
3 CHOH*	3 CHOH*	3 CHOH*	3 CHOH*
4 CHOH*	4 CHOH*	4 CHOH*	4 CHOH*
5 CH₂OH	5 CHOH*	5 CH₂OH	5 CHOH*
Aldopentose	6 CH₂OH	**Ketopentose**	6 CH₂OH
	Aldohexose		**Ketohexose**

Carbon atoms not situated at the ends of the molecule always have four different substituents and thus represent stereogenic centers. Such chiral C-atoms are labeled with an asterisks in the scheme above. Since every chiral carbon atom can have two configurations, every monosaccharide can have 2^n stereoisomers with n being the number of stereogenic centers. For instance, aldohexose can have 16 stereoisomers. This is a good example of a large variety of possible structures for organic compounds. As we recall from the chemistry of amino acids, nature is very selective in the generation of specific isomers. The most common monosaccharide found in living organisms is the aldohexose traditionally called **D-glucose**.

The configurations of all chiral C-atoms in the glucose molecule are shown in the formulas of the scheme below. Here, the structures A and B represent a glucose molecule using the wedge-dash notation, C is the Fischer projection formula and D is the shorthand Fischer formula. The horizontal lines label the positions of -OH groups. Remember that the substituents written on the left and right side of the vertical line are located above the plane of drawing.

| A | B | C | D |

D-Glucose

Let us give some more examples of monosaccharides. **D-mannose** belongs to the aldohexoses, **D-ribose** and **D-arabinose** belong to the aldopentoses and **D-fructose** is the most important ketohexose. The label D represents the relative configuration in accordance with the rules which were discussed in the section on amino acids. Since the monosaccharide molecule possesses several stereogenic C-atoms the labels D or L are used only for the chiral carbon that is furthest from the aldehyde or the ketone group. Where all natural amino acids have the L-configuration, all natural monosaccharides have the **D-configuration**. This is another example of selectivity in nature!

D-Mannose

D-Ribose

D-Arabinose

D-Fructose

The evidence that in natural products, especially in primary metabolites, some structural patterns are repeated in different structures, implies that most organisms have common biosynthetic pathways. At the end of the nineteenth century when the biosynthetic pathways were not known, **Heinrich Kiliani** and **Hermann Emil Fischer** developed a simple model which could explain how stereochemically pure long-chained monosaccharides can form from simple molecules. The simplest, parent monosaccharide is **aldotriose** with only one chiral carbon atom. Hence, there are D- and L-aldotrioses. Since aldotriose is the product of mild oxidation of glycerol it can also be called **glyceraldehyde**.

CH₂——OH   CHO   CHO

CH——OH   OXIDATION   H———OH   HO———H

CH₂——OH   CH₂OH   CH₂OH

Glycerol   **D-Glyceraldehyde**   **L-Glyceraldehyde**
                            (D-Aldotriose)   **(L-Aldotriose)**

To extend a molecule of aldotriose (glyceraldehyde) by an extra carbon atom, Kiliani and Fischer applied the already known reaction of nucleophilic addition to the aldehyde group. They used the cyanide ion ⁻CN as the nucleophile. After the addition of cyanide to the aldehyde group, whose carbon atom is achiral, the carbon atom becomes chiral in the product because of the four different substituents attached. This change in stereochemistry is shown in blue in the following scheme. The products are two diastereomers in which the new stereogenic center (the C-atom bound to the CN group) can have the D- or the L-configuration.

CHO   CN   CN

H———OH   :CN, H⁺   H———OH   HO———H

                                H———OH   +   H———OH

CH₂OH   CH₂OH   CH₂OH

D-Glyceraldehyde
(D-Aldotriose)

Since the aldehyde carbon is bound to oxygen by the double bond this carbon is coplanar with all three substituents. The ⁻CN nucleophile can then attack this carbonyl plane from either side with equal probability which leads to the formation of two products that differ in their relative configurations, D or L (scheme below).

If the cyanhydrine products are then first oxidized to acids and subsequently reduced to aldehydes, the final products are two new monosaccharides, the aldotetroses **D-erithrose** and **D-threose**:

D-Erythrose

D-Threose

This **Kiliani-Fischer synthesis** can be continued starting from these new tetroses (D-erithrose and D-threose) to prepare four new monosaccharides (aldopentoses), because every tetrose can yield two new pentose diastereomers. By repeating this series of reactions it is possible to extend the carbon chain further to get hexoses. If the synthesis starts with D-glyceraldehyde all the product molecules of the corresponding aldoses will have the D-configuration on the C-atom which is furthest from the aldehyde group. These monosaccharides belong to the **D-series**. On the other hand, aldoses prepared from L-glyceraldehyde will have the last C-atom with the L-configuration. These monosaccharides belong to the **L-series**. All living organisms produce only the D-series of carbohydrates!

11.2.1 Cyclic Structures of Monosaccharides

The representation of monosaccharide molecules as open-chain structures is only partially correct. In solution, the molecules with five- and six carbon atoms can also appear in cyclic forms which are in equilibrium with the open-chain structures. The cyclization process represents the intramolecular nucleophilic attack of the hydroxyl group on the aldehyde or keto group.

D-Glucose

D-Glucose
(another conformation)

α-D-Glucopyranose

Pyrane

β-D-Glucopyranose

The carbon atom, which in the open-chain form was part of the carbonyl group, becomes chiral after cyclization and is called the **anomeric carbon atom**. In the cyclic form the anomeric carbon can have two configurations which are labeled with the Greek letters α and β. Stereoisomers that differ in the configuration at this new chiral center are called **anomers**. For example, the cyclic form of D-glucose can have α and β anomers named **α-D-glucopyranose**, and **β-D-glucopyranose**. The word **pyranose** has been taken from the analogous cyclic ether with the six-membered ring which is known as **pyrane**. In the first chapter of this book we have claimed that the substituted cyclohexane ring is most stable when its substituents are in equatorial positions. The anomeric carbon atom is an exception; the most stable anomer has the OH group in the axial position which makes it the α-anomer. This is called the anomeric effect.

Anomers also appear in cyclic form as five-membered rings, for instance in the monosaccharides fructose and ribose. These cyclic molecules are named **furanoses** because of their similarities to the cyclic ether **furan**. Starting from D-fructose and D-ribose, the cyclization yields anomers **α-** and **β-D-fructofuranose**, as well as **α-** and **β-D-ribofuranose**, respectively.

D-Froctose   **α-D-Fructofuranose**   **β-D-Fructofuranose**

Furan

D-Ribose   **α-D-Ribofuranose**   **β-D-Ribofuranose**

In analogy with pyranoses, the furanoses are also in equilibrium with the open-chain forms. The new functionality which appears in the cyclic forms is **hemiacetal**, the structure already mentioned in our discussions on alcohol addition to the carbonyl group. The hemiacetals can easily undergo hydrolysis yielding the open chain monosaccharide molecule.

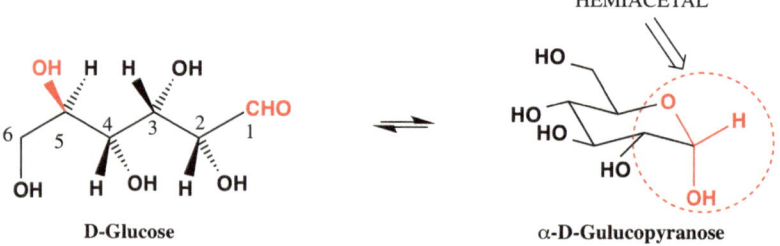

D-Glucose   **α-D-Gulucopyranose**

Representation of the cyclic forms could be simplified by using the special short-hand notation called the Haworth formula (in the honor of **Sir Norman Haworth**). Fischer formulas can be easily transformed into Haworth formulas by the rules given in the following scheme.

D-Glucose   D-Glucopyranose   α-D-Glucopyranose

D-Fructose   D-Fructofuranose   α-D-Fructofuranose

FISCHER`S FORMULA            HAWORTH`S FORMULA

Chiral carbon atoms that do not contribute to the formation of the cyclic form have substituents attached labeled in blue. In D-glucopyranose these atoms are numbered as 2, 3 and 4 while in D-fructofuranose these atoms are 3 and 4. The rule for writing Haworth formulas says that the substituents **below the plane of the ring in Haworth projections are equivalent to those on the right-hand side of the corresponding Fischer projection** (shown in blue).

11.2.2   Disaccharides and Polysaccharides

The hydroxyl group on the anomeric carbon of the monosaccharide molecule can be replaced by another monosaccharide molecule. The resulting dimer is called a **disaccharide**. Binding of α-D-glucopyranose to β-D-fructofuranose gives the disaccharide α-D-glucopyranosyl-β-D-fructofuranoside, the disaccharide known under the name **sucrose** which is commonly used in food preparation (see the scheme). Analogously, β-D-galactopyranose can bind with β-D-glucopyranose into the disaccharide β-D-galactopyranosyl-β-D-glucopyranoside or **lactose**, the main sugar component of milk. The chemical bond by which the monosaccharide units are joined is called the **glycosidic bond**. In lactose, carbon atom 1 on one monosaccharide is bonded to carbon 4 of the another monosaccharide and such bond is known as the **1,4 glycosidic bond**. This is the most frequent mode of binding between monosaccharide units.

α-D-Glucopyranose   β-D-Fructofuranose   α-D-Glucopyranosyl β-D-fructofuranozide
(Sucrose)

β-D-Galactopyranose   β-D-Glucopyranose   β-D-Galactopyranosyl β-D-glucopyranoside
(Lactose)

In the same way, glucose molecules can be connected by a glycosidic bond into long polymeric structures of **polysaccharides**. The structure of the polysaccharide formed from glucose depends mostly on the starting anomer. Polymerization of α-D-glucopyranose by the α(1,4)-glycosidic bond yields **amylose**, the polysaccharide which is a component of **starch**.

α-**D-Glucopyranose**

Amylose

Polymerization of another anomer, β-D-glucopyranose by the β(1,4)-glycosidic bond forms **cellulose**.

B-**D-Glucopyranose**

Cellulose

11.3   Glycosides and Nucleotides

Glycosides are compounds in which the substituent is bound to the anomeric carbon. This group comprises numerous natural products; the primary metabolites. The most important are **nucleosides**, the basic structural units from which **nucleotides** and **nucleic acids** are formed. The monosaccharide component of the nucleoside is **D-ribose** or its partially reduced form **D-deoxyribose**. In biochemistry and molecular biology the most important are the nucleosides in which ribose is bonded to heterocyclic molecules called **nucleic bases**. The most common nucleic bases are **pyrimidines** and **purines**.

| Pyrimidine | Uracil | Thymine | Cytosine |

PYRIMIDINE BASES

Pyrimidines are derivatives of the cyclic amine **pyrimidine** while purines are derived from the bicyclic amine **purine**.

Purine   Adenine   Guanine

PURINE BASES

The glycoside that consists of adenine and D-ribose (in its cyclic form ribofuranose) is **adenosine**.

Adenine fragment

Ribofuranose fragment

Adenosine

In biochemical systems, adenosine appears as a phosphate in three forms, as **adenosine monophosphate (AMP), adenosine diphosphate (ADP)** and **adenosine triphosphate (ATP)**. These phosphate esters of nucleosides are called nucleotides and they are the basic building blocks of nucleic acids **DNA** and **RNA**, the compounds which are carriers of genetic information.

Adenosine monophosphate
AMP

Adenosine diphosphate
ADP

Adenosine triphosphate
ATP

In an enzyme catalyzed reaction, ATP can be hydrolyzed by losing one phosphate group PO_4^{3-}, with the release of large amount of energy that has been accumulated in the P-O bond. This reaction serves as the source of energy for the processes in living cells.

Nucleic acids are polymers made from nucleotides. In living organisms two main sorts of nucleic acids are present: **ribonucleic acid, RNA** and the **deoxyribonucleic acid, DNA**. The building blocks of nucleic acid molecules comprise five nucleotides bound either to the **ribofuranose** (as in RNA) or to the **deoxyribofuranose** (as in DNA) and to the phosphate group. Nucleotides which are phosphates of adenine, thymine, guanine and cytosine are the components of DNA. In RNA the thymine unit is replaced with uracil. Polymeric molecules of nucleic acids can exist as either **single-stranded** or **double-stranded** chains. In the single-stranded chain the ribose molecules are interconnected via the phosphate groups. The structure of the nucleic acid chain is represented in the scheme below. RNA and DNA differ in the substituent Z on C_2 carbon atom of the monosaccharide unit. If Z is the

OH group, the fragment of the nucleic acid molecule would represent RNA, but if Z is the hydrogen atom, the structure would represent DNA. Ribose and deoxyribose differ in this detail only.

Z = OH ribose, RNA

Z = H deoxyribose, DNA

β-D-Ribofuranose
Ribose

β-D-Deoxyribofuranose
Deoxyribose

The double-stranded DNA consists of two parallel chains folded into an α-helix. The chains are joined by hydrogen bonds that follow the **rules of complementarity**. In accordance with these rules, adenine is always bound with thymine of the other chain and guanine is always bound to cytosine (see the following scheme). The principles of molecular complementarity do not apply only to nucleic acids. Rather, they represent general rules for the formation of various **supramolecular structures**. If nucleotides are labeled by their first letters, the complementary pairs would be AT and GC, respectively. The structural key for complementarity is the number of hydrogen bonds. There are two such bonds in the AT pair and three in

the GC pair. For example, it is impossible for G to be bound to T: the nucleotides
are **selective**.

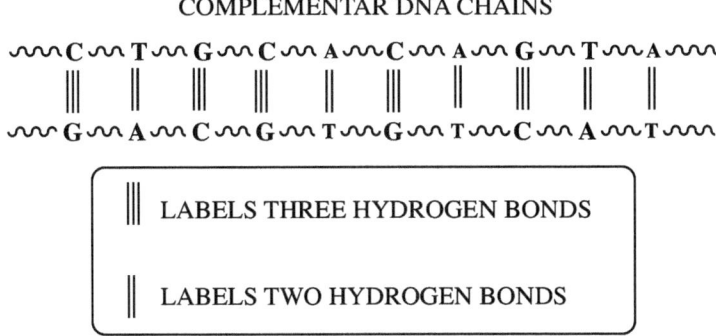

Based on the complementarity principle, two DNA chains bind as shown in the
scheme:

COMPLEMENTAR DNA CHAINS

In biochemical processes, the DNA molecule can be separated into two single-
stranded molecules. One of such single-stranded intermediates serves as a matrix on
which the complementary chain can be synthesized. This is the main principle of
DNA **replication**. The mechanism and numerous details of this process are studied
in biochemistry.

11.4   Lipids

At the beginning of this Chapter we have mentioned that one of the oldest methods for isolation of organic natural products is extraction. Primary metabolites discussed up to now were mainly ionic or polar compounds soluble in water or in polar solvents. Organic natural products that belong to different structural classes but which have in common the solubility in nonpolar solvents are called **lipids**. Some of the lipids are esters of long-chained fatty acids and they can be readily hydrolyzed under acidic or basic conditions. In this group we encounter **waxes**, **fats** and **phospholipids**.

11.4.1   Waxes

Waxes are structurally the simplest lipids, being esters of long-chained carboxylic acids and long-chained alcohols. The chains can have 12 or more carbon atoms. The ester represented in the following scheme is the main component of wax produced by bees. It can be easily saponified into the corresponding long-chained acid and alcohol.

$$CH_3(CH_2)_{14}\overset{\overset{\displaystyle O}{\|}}{C}OCH_2(CH_2)_{28}CH_3$$

$$\downarrow \quad NaOH, H_2O$$

$$CH_3(CH_2)_{14}CO\overset{-}{O}\ Na^+ \quad + \quad HOCH_2(CH_2)_{28}CH_3$$

The reaction of saponification has already been discussed in the chapter on esters. This reaction is the basic procedure for decomposition of ester lipids.

11.4.2   Fats

Fats are esters of long-chained fatty acids and the alcohol glycerol. Since glycerol has three hydroxyl groups its complete esterification requires three fatty acid molecules. The chains in the molecules of fatty acids can be saturated or partially non-saturated. Some examples are given in the following table.

$CH_3(CH_2)_{10}COOH$	Lauric acid
$CH_3(CH_2)_{12}COOH$	Myristic acid
$CH_3(CH_2)_{14}COOH$	Palmitic acid
$CH_3(CH_2)_{16}COOH$	Stearic acid
$CH_3(CH_2)_7CH=CH(CH_2)_7COOH$	Oleic acid
$CH_3CH_2CH=CHCH_2CH=CHCH_2CH=CH(CH_2)_7COOH$	Linolenic acid

Fats and oils differ according to the structures of their fatty acid. While fats are made of saturated acids, oils are esters of unsaturated fatty acids. As we know from the chemistry of hydrocarbons, unsaturated compounds such as alkenes or alkynes can be saturated by addition of hydrogen in a metal-catalysed reaction. Using the same method, natural oils can be transformed to fats that are frequently called synthetic fats. The process of hydrogenation which yields the fat tristearin is represented in the following scheme.

1,2-Dioleyl-3-stearylglycerol

H_2, Pd - catalyst

Tristearyl glycerol
(Tristearine)

Because of the presence of double bonds, the digestion of oils is easier than the digestion of saturated fats. A recent nutritional recommendation is to use oils that consist of **ω-3 fatty acids**. Another name for them is ***n*-3 fatty acids**. These compounds have hydrocarbon chains consisting of 18–22 carbon atoms in which the first of several double bonds resides on the third C-atom counting from the alkyl (not carboxyl!) terminus of the molecule. Linoleic acid is a representative example of ω-6 fatty acids. In addition, the molecules of unsaturated fatty acids have a Z-configuration around the carbon-carbon double bonds!

Saponification of fats with KOH or NaOH yields **soaps**, the sodium or potassium salts of fatty acids and glycerol. The behavior of soaps in water solutions and the formation of micelles have already been discussed.

Tristearylglycerol
(Tristearin)

KOH

Potassium stearate
(soap)

Let us mention the old tradition known from ethnology for the preparation of soaps by boiling animal fat with the ash that remains after the burning of beech wood. This ash contains potassium carbonate and hydroxide which serve as bases for the saponification. Interestingly, in the Arabic language, the word for ash is *kali*. The terms *alkali* or *kalium* (for the element potassium) were derived from this root. The arabic culture and tradition strongly influenced European alchemy and its transformation into modern chemistry.

11.4.3   *Phospholipids*

The property of the glycerol-based lipids to form micelles even without saponification is especially pronounced in phospholipids. The basic structural unit of the phospholipid molecule is **phosphatidic acid** shown in the scheme below. The substituents R and R′ are long-chained alkane or alkene groups. The phosphate group can be bound in different structures such as **choline** or **serine**, molecules which play important roles in metabolism.

Glycerol   Phosphatidic acid

Choline

Serine

Phosphatidylserine

Phosphatidylcoline
(Lecithin)

Compounds similar to **lecithin** are components of cellular membranes because they can form double-layered structures in which the hydrophobic sides are oriented towards each other. Being hydrophilic, the choline group is oriented towards the water solvent.

1,2-Distearylphosphatdiyil choline

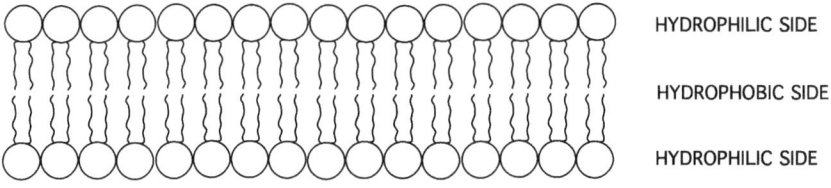

DOUBLE-LAYER STRUCTURE OF CELLULAR MEMBRANE

Such membranes prevent the passage of Na^+ and K^+ ions, the process important for the transmission of electrical signals between lining cells. Being hydrophilic, these ions cannot pass through the hydrophobic layer of the membrane. For ion transport through membranes, the cells have developed special catalytic systems which require specific groups of enzymes.

11.4.4 Terpenes and Steroids

In contrast to fats and phospholipids, terpenes and steroids cannot be hydrolyzed to simpler units. However, **Otto Wallach** in the nineteenth century and **Leopold Ružička** (Croatian Nobel laureate) in the first half of twentieth century discovered that these compounds can be decomposed into simpler structures that consist of five carbon atoms. These units are structural analogs of the alkene named **isoprene** (2-methylbuta-1,3-diene). Isoprene can be regarded as a molecule containing a head and a tail.

HEAD   TAIL

Isoprene

Isoprenic units can be interconnected in different ways: head-to-head, tail-to-tail, head-to-tail or tail-to-head. Oligomers of such isoprene units are called **terpenes**. A molecule with two isoprenes is named **monoterpene**, a molecule with four units is **diterpene**. Three isoprene monomers form **sesquiterpenes**. One of the most important terpenes is **squalene**, a primary metabolite that is present in all living organisms. In contact with the atmospheric oxygen, squalene is oxidized to **squalene-oxide**. In the reaction with the corresponding enzyme, squalene-oxide is folded into the conformation represented in the scheme below.

Squalene

$\downarrow$ **O$_2$**

Squalene oxide

In the subsequent reaction step, the epoxide ring is protonated to a carbocation which activates successive cyclizations of four double bonds to give the tetracyclic structure of **lanosterol**.

H$^+$

HO

Lanosterol

HO

**CONFIGURATION
of
Lanosterol**

Lanosterol is the precursor for the biosyntheses of primary metabolites that have a common structural motif of three 6-membered rings and one 5-membered ring. These compounds are known as **steroids**. In the chemistry of natural products the rings are labeled by capital letters A, B, C and D.

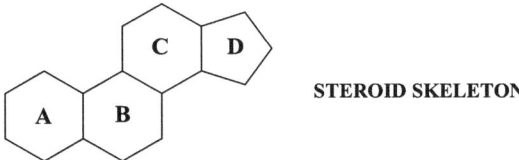

STEROID SKELETON

Some steroids, such as **cholic acid**, **progesterone** and **testosterone** were already mentioned in the chapters discussing aldehydes, ketones and carboxylic acids. The most common steroid in humans is **cholesterol**. Although the compound has been discovered in the eighteenth century its complete molecular structure was determined only in the middle of the last century. Cholesterol appears in most of tissues and it has a special role in the regulation of blood circulation. An imbalance of cholesterol in the organism can cause serious health problems similar to artherosclerosis. The cholesterol molecule, like other steroids, is formed by a particular biosynthetic pathway from the terpene precursors, squalene and lanosterol. Since cholesterol has 27 carbon atoms, 3 atoms less than the triterpene squalene (which has 30 C-atoms), 3 C-atoms are eliminated during the biosynthetic process.

Cholesterol

Under exposure to sunlight the cholesterol in the skin is oxidized to 7-dehydrocholesterol, which is immediately rearranged to **vitamin D$_3$**.

Cholesterol

OXIDATION
ENZYME
- H$_2$

7-Dehydrocholesterol

LIGHT

Vitamin D$_3$

Most hormones are also steroids. Besides the already mentioned sexual hormones testosterone and progesterone, there is also **estradiol**, the female sexual hormone, which is responsible for the development of the secondary characteristics of the female anatomy.

Estradiol

11.4.4.1   Terpenes as Secondary Metabolites

We have already mentioned that secondary metabolites are compounds which are specific to particular biological species. They are not common to all living organisms. Terpenes, which were discussed in the previous section, are primary metabolites. However, many organisms produce secondary metabolites structurally similar to terpenes, i.e. their molecules can be decomposed into isoprene units. One of them is the already mentioned camphor but more examples are given below.

Formulas of terpenes in the scheme above are presented in two rows. In the bottom row, if the chemical bonds labeled in red are deleted the residues represent isoprene units. Based on the number of isoprene fragments, **menthol** is monoterpene, **α-selinene** belongs to sesquiterpenes and **cembrene** as well as **vitamin A** are diterpenes. Menthol is the component of peppermint responsible for its refreshing odor. The aroma of celery originates from **α-selinene** while **cembrene** can be isolated from pine trees. The derivative of **retinal** has already been discussed in the section about the *cis-trans* isomerism in the biochemical process of vision.

11.5   Alkaloids

The first secondary metabolites that were isolated from natural sources are the alkaloids, these were also the compounds which were the first group of natural products studied systematically by chemists. These compounds were named **alkaloids** because all of them have nitrogen in their molecules and it is known that nitrogen causes alkaline behavior. As is typical for all secondary metabolites, the molecules of alkaloids can have a wide variety of structures. Many alkaloids induce marked physiological effects, especially on the nervous systems. Some of them like **nicotine** or components of opium are drugs or have analgesic properties.

Adrenaline is produced in the brain as a stress activator and it causes increased psychophysical activity. Compounds isolated from special sorts of poppy have frequently been used, in the form of impure substances, as tranquilizers and are known under the name **opium**. The name is drawn from the Greek word οπιον which means poppy. The first pure substance isolated from opium is **heroin**, a compound that has also been synthesized for medicinal use.

In everyday use as pain killer, the acetylsalicylic acid can be isolated from the willow bark (*salix*). On the market this compound, which is not an alkaloid, is known under the name **aspirin**. Many antibiotics are also alkaloids. One of the alkaloids, **quinine** is very efficient drug against malaria.

Nicotine

Adrenaline

Opium alcaloids

R = R′ = H   Morphine

R = R′ = $OCCH_3$ Heroine

R = CH_3 R′ = H   Codeine

Acetylsalicylic acid
(Aspirin)

Quinine

11.6   Organic and Bioorganic Reactions

Most of the reactions mentioned so far are used in laboratory preparations. The difference between bioorganic and laboratory synthesis has already been mentioned in the context of the synthesis and biosynthesis of amino acids. The Strecker synthesis was compared with reactions catalyzed by the enzyme aminotransferase. The most important difference between the laboratory synthesis and biosynthesis is that biosynthesis is strongly stereospecific. In living cells the reaction conditions are fundamentally different from those in the organic laboratory. In the laboratory, the reactions are performed under drastic conditions of elevated temperature or pressure, frequently by using strong reagents such as metal catalysts and/or strong acids and bases. Besides, most of the reactions in the laboratory occur in organic solvents and within a wide range of acidity. In contrast, the living cell is made mostly from water and is very sensitive to strong inorganic reagents. In most cells the acidity is nearly constant with the pH maintained close to the value of 7.5.

For the reactions such as nucleophilic substitutions, eliminations or redox reactions, living systems do not only use special catalysts such as enzymes, but also specific biologically compatible reagents, leaving groups and nucleophiles.

As we know, nucleophilic substitution can occur via two possible mechanisms, S_N1 or S_N2. While the S_N2 mechanism depends on the concentration and on the nature of the nucleophile, the S_N1 mechanism depends only on the nature of the leaving group. The reaction intermediates in S_N1 reactions, the carbocations, are the molecules that can be effectively stabilized by the enzyme catalysts and the reaction conditions for S_N1 processes can be milder. Consequently, most nucleophilic substitutions in living cells occur via the S_N1 mechanism. For the interconversion of functional groups by nucleophilic substitution in the laboratory we use alkyl halides as a starting materials. Since alkyl halides are hydrophobic, they cannot persist in the aqueous medium of the living cell and halide as a leaving group is not compatible with bioorganic systems. Leaving groups in cells must be hydrophilic such as **pyrophosphate** esters, which are ionic molecules well soluble in water.

Carboxylic acid derivatives are prepared in the laboratory starting from acyl chlorides. Since acyl chloride reacts vigorously with water, yielding HCl, using it would kill the cell. Hence, the leaving group in the biosynthesis of derivates of carboxylic acids, **coenzyme A** has a special and more complex structure.

Oxidations and reductions can be performed in the laboratory by using metallates or metal hydrides. Under very mild conditions, living cells use special redox systems which involve enzymes and the corresponding active reagents with complex structures.

The most common reagents are **nicotinamide adenine dinucleotide NAD⁺** and its phosphate derivative **NADP⁺**. In reduced form these molecules are transformed to **NADH** and **NADPH** by accepting hydrogen atom.

Nicotinamide adenine dinucleotide
oxidised form NAD
for R = H
Nicotinamide adenine dinucleotide phosphate
oxidised form NADP
for R = PO$_3^{2-}$

Nicotinamide adenine dinucleotide
reduced form NADH
for R = H
Nicotinamide adenine dinucleotide phosphate
reduced form NADPH
for R = PO$_3^{2-}$

NAD⁺ is efficient in reactions which yield functional groups with double bonds between a carbon atom and the heteroatom. In this way alcohols can be converted to ketones. In the laboratory we need strong oxidants such as $KMnO_4$ or $K_2Cr_2O_7$ to oxidize alcohols.

For the reduction of alkenes or alkynes to alkanes in laboratory we use metal catalysts such as Pt or Pd and often high pressures. The heating of alkane precursors with these metal catalysts reoxidizes alkanes to alkenes. In biosynthesis these reactions proceed with special reagents like **flavine adenine dinucleotide FAD** or its reduced form **FADH₂**.

Flavin adenine dinucleotide oxidised form FAD

$-H_2$ $+H_2$

Flavin adenine dinucleotide reduced form FADH$_2$

An example of reactions mediated by FAD or FADH$_2$ is given in the scheme:

One of the most important redox processes in living organisms includes the transfer of oxygen in erythrocytes by the protein **hemoglobin**. The active functionality in hemoglobin is an organometallic group called **heme**. The organic part of this complex is the **porphyrin** aromatic heterocyclic molecule (**heterocyclic structures are cyclic molecules with heteroatoms in the ring**).

Porphyrin

Heme

HOOCCH$_2$CH$_2$ CH$_2$CH$_2$COOH

The oxygen molecule binds to the iron atom situated in the molecular plane of the porphyrin ring. The opposite side of the plane is occupied by the **histidine** amino acid that is part of the polypeptide chain of the protein molecule.

The presence of the protein is of high importance because without it the oxygen atom close to the iron atom would immediately oxidize Fe^{2+} in heme to Fe^{3+} and the oxygen molecule would then not bind to iron. In addition, the iron ion in the complex with porphyrin, oxygen and histidine forms an octahedral coordination, a sterically favored structure.

Heme

Histidine from protein polypeptide chain

Heme - oxygen complex

In this comparison between the reactions in the living organisms and those in the laboratory, we have discussed only the most common examples. We did not mention the most important property of bioorganic reaction systems, their interconnectivity. Biochemical reactions form organized systems that act more or less autonomously.

Index

A

Absolute configuration, 108, 136
(R) Absolute configuration, 109
(S) Absolute configuration, 109
Absolute methanol, 71, 72
Acetal, 90
Acetaldehyde, 87, 89, 137
Acetamide, 117
Acetanhydride, 112
Acetic acid *(acidum aceticum)*, 96, 97, 100,
    112, 113, 117
Acetone, 31, 87, 88, 95
Acetylene, 39
Acid halides, 111
Acids, 50, 72–75, 77, 82, 90–92, 95–100,
    103, 111–115, 119, 121, 136, 147,
    158, 166
Acidum aceticum, 96
Acidum formicum, 96
Acrolein, 88
Activation (of benzene ring), 122, 124
Activation energy, 67–69
Acylation, 120, 121
Acyl group, 111, 120
Adamantane, 16
Addition, 46–51, 56, 82, 86, 94, 95, 106, 120,
    129, 146, 150, 159
Adenine, 154–157
Adenosine, 154, 155
Adenosine diphosphate (ADP), 155
Adenozine monophosphate (AMP), 155
Adenozine triphosphate (ATP), 155
Adipic acid, 117
ADP. *See* Adenosine diphosphate (ADP)
Adrenaline, 165
Alanine, Ala, 135–137
Alcoholates, 71, 73, 76, 90

Alcohols, 18, 31, 65, 70–78, 80, 81, 86, 90,
    92–95, 100, 112–114, 116, 150,
    158, 168
Aldehyde, 18, 31, 51, 52, 86–95, 114, 136,
    144–148, 163
Alder, K., 129
Aldohexose, 144, 145
Aldol, 92, 93
Aldol condensation, 92, 93
Aldopentose, 145, 148
Aldose, 144, 148
Aldotriose, 146
Alkaloids, 131, 165–166
Alkanes, 1–17, 30, 39, 46, 117,
    160, 168
Alkenes, 1, 18, 30–31, 39–57, 75, 77, 108,
    119, 129, 140, 159–161, 168
Alkoxides, 76, 111
Alkylation, 120
Alkyl halides, 47, 60–63, 65, 76, 77, 80, 94,
    96, 120, 167
Alkynes, 1, 18, 31, 39–57, 86, 159, 168
Allene, 40
Amides, 19, 111, 115–118, 138, 139
Amines, 19, 80–83, 94, 111, 116, 117, 154
Amino acids, 131–145, 166, 170
α-Amino acids, 132
β-Amino acids, 132
γ-Amino acids, 132
Aminotransferase, 137, 138, 166
Ammonia, 22, 80–82, 116
Ammonium cyanate, xi, 2
AMP. *See* Adenozine monophosphate (AMP)
Amylose, 152, 153
Anhydride, 111–112
Aniline, 82, 83, 117, 122–124
Anionic micelle, 101

H. Vančik, *Basic Organic Chemistry for the Life Sciences*,
DOI 10.1007/978-3-319-07605-8, © Springer International Publishing Switzerland 2014